Louis Figuier, S. R. Crocker

The To-morrow of Death

The Future Life According to Science

Louis Figuier, S. R. Crocker

The To-morrow of Death
The Future Life According to Science

ISBN/EAN: 9783337404529

Printed in Europe, USA, Canada, Australia, Japan

Cover: Foto ©berggeist007 / pixelio.de

More available books at **www.hansebooks.com**

THE

TO-MORROW OF DEATH;

OR,

THE FUTURE LIFE ACCORDING TO SCIENCE.

BY LOUIS FIGUIER,

AUTHOR OF "PRIMITIVE MAN," "EARTH AND SEA," ETC.

TRANSLATED FROM THE FRENCH BY S. R. CROCKER.

NOTE BY THE TRANSLATOR.

In the ensuing translation, a literal rendering of the original has been aimed at, as preferable, considering the gravity of the general question discussed, and the purely scientific methods with which it is treated, to a more elegant and less faithful paraphrase. The peculiarities of the author's style have been respected in this version, in order that the reader might have the fullest possible information on which to base a judgment of the work.

Boston, January, 1872.

CONTENTS.

CHAPTER VIII.

CHAPTER IX.

CHAPTER X.

CHAPTER XI.

CHAPTER XII.

CHAPTER XIII.

CHAPTER XIV.

Page

What are the Relations that subsist between Ourselves and the Superhumans? 177

CHAPTER XV.

What is the Animal? The Soul of Animals. The Migration of Souls through the Bodies of Animals 192

CHAPTER XVI.

What is the Plant? The Plant feels. How Difficult it is to distinguish Plants from Animals. The General Chain of Living Beings 202

CHAPTER XVII.

Proofs of the Plurality of Human Existences and Incarnations. Outside of this Doctrine we can explain neither the Presence of Man on the Earth, nor the Painful and Unequal Conditions of Human Life, nor the Fate of Children who die in Infancy . 229

CHAPTER XVIII.

The Faculties peculiar to some Children and the Aptitudes and Natural Vocations among Men constitute other Proofs of Reincarnations. Explanation of Phrenology. The "Innate Ideas" of Locke, and Dugald Stewart's "Principle of Causality," are explicable only on the Hypothesis of a Plurality of Lives. Vague Recollections of Former Existences 242

CHAPTER XIX.

The Hypothesis of Successive Existences compared with Materialism, and with the Dogmatic Christian View of the Destiny of Man. Punishments and Rewards in the Christian Scheme, and in the Doctrine of Successive Existences 256

CHAPTER XX.

Summing up of the System of the Plurality of Existences . . . 273

CHAPTER XXI.

CHAPTER XXII.

CHAPTER XXIII.

THE TO-MORROW OF DEATH.

INTRODUCTION.

READER, you must die. Perhaps you will die to-morrow. What is going to happen to you, and what will you be on the to-morrow of your death? I speak not of your body: that will be of no more account than the vestments that cover it, or the shroud which will enwrap your remains. Like these vestments, like the funeral sheet with which you will be covered, your body will dissolve, and its elements will lose themselves in the great material reservoirs of Nature, — in the air, in the earth, in the water. But your soul, whither will that go? That which in you has felt, has loved, has suffered, has been free, — what will become of it on the to-morrow of your death? You surely will not contend that your soul, as well as your life, will be annihilated on the day of your death, and that nothing will be left of that which beat in your bosom, which vibrated to emotions of happiness or of grief, to the gentle affections, the thousand passions and agitations of life. But where will that thinking soul go, which must endure beyond the tomb? what will become of it? and what will you be, O reader! on the to-morrow of your death?

Such is the question which it is attempted to fathom in this book.

Almost all thinkers have declared the problem of human life to be insoluble. They have maintained that the human intellect is powerless to pierce so profound a mystery, and that, in a case like this, the only course is to forbear. Most men, through either carelessness or conviction, reason in this way. Moreover, when one dares to look for a moment in the face of so tremendous a question, he finds himself instantly surrounded by darkness so black that he has no courage to prosecute his inquiry. Thus we are led to shrink from all thoughts of the future life.

There are nevertheless emergencies in which we are compelled to meditate on this solemn and difficult subject. When we find ourselves in danger of death, or when we have lost some dearly loved friend, we cannot help thinking of the future life. After having thoroughly examined this idea, we may be led to realize that the problem is not so far elevated beyond the reach of human thought as we had believed.

During the greater part of his life, the author of this book had believed, like all the world, that the problem of the future life was beyond our mental grasp, and that it was the part of wisdom not to trouble one's mind with it. But one day — one dreadful day! — a thunderbolt struck him. He lost his beloved son, on whom all the hopes and ambitions of his life were centred. Then, and in the bitterness of his grief, he pondered on the new life that must open to us beyond the grave. Having dwelt long on this idea in his solitary musings, he asked of the exact sciences what positive evidence they could render on this point. Finally he questioned ignorant and simple men, country peasants, and the unlettered crowds of cities, — an always valuable source of information to him who seeks to repair to the true principles of Nature; for Nature is changed neither by the prejudices of education nor by the routine of a hackneyed philosophy.

Thus the author was led to form a whole system of ideas about this new life which will begin for mankind after this earthly sojourn.

But all is included in Nature. Every organized existence is bound to another which precedes it, and to another which follows it, in the scale of living creatures. The plant and the animal, the animal and the man, are connected, united, one with the other. The natural order and the moral order are often identical. Thus it follows that he who thinks he has discovered the explanation of some single fact, in the organization, is soon led to extend this explanation over the whole kindred of living existences, and to ascend, round by round, the whole ladder of Nature. Such was the author's experience. Having sought to learn what becomes of man when he leaves this earthly life, he was led to apply his views to all living beings, to animals and then to plants. The force of logic drove him to cover with his system the beings, invisible to our eyes, who must inhabit the planets, the sun, and all the innumerable stars scattered over the vast expanse of the heavens. Hence this work is not only an attempt to solve the problem of the future life by scientific methods, but also a statement of a complete theory of Nature, a real philosophy of the universe.

I may be deceived; I may mistake for serious opinions the dreams of my imagination; I may lose myself in the dark region through which I try to grope my way: but I write in absolute sincerity; that is my excuse. I hope, moreover, that my example will incite others to make a like attempt, — to apply the exact sciences to the study of the great question of man's fate after death. A series of essays in this direction would be of incalculable service to natural philosophy as well as to human progress.

After those terrible hecatombs, which in 1870 and 1871 drenched in blood our unhappy land, there is not a family in France which has not to lament a lost relative or friend. Having found, not consolation for my grief, but relief for my mind, in the composition of this work, I have thought that those who suffer and those who weep would feel, in reading these pages, the same sentiments of hopefulness which have soothed my saddened heart.

Society is a prey to a terrible evil, a moral cancer which threatens to destroy it: it is materialism. Advanced first in the universities and scientific literature of Germany, materialism has spread over all France. It has lost no time in descending from the desks of *savants* among the more enlightened classes, and soon among the people. And this people has undertaken to show us the practical results of materialism. Little by little it has freed itself from every restraint, cast off every moral obligation: it accepts no more neither religion nor its ministers, nor social hierarchy, nor country, nor liberty. All this was sure to have a dangerous end. After a long political anarchy, these furious fools have led through the capital of France terror, flames, and death.

It is not petroleum which set fire to the monuments of Paris: it is materialism. It is plain enough that the moment one is convinced that all is finished on earth, that there is nothing after this life, we have only, one and all of us, to appeal to violence, to provoke disturbance and anarchy everywhere, to find in this propitious disorder the means of satisfying our brutal desires, our ambition, and our sensual passions. Civilization, society, and morals are like a string of beads, whose knot is belief in the immortality of the soul:

break the knot, and the beads scatter. Materialism, then, is the parent of all the evils of European society: it is the plague of our day. Now materialism is attacked-in-breach in this book, which may be called Spiritualism demonstrated by Science. It is especially this last consideration that the author's friends have effectively employed to determine him to bring out this work without delay.

Such are the motives that have induced me to publish "The To-morrow of Death;" and, these having been stated, I enter upon the work.

CHAPTER I.

Man is the Result of the Triple Alliance of the Body, the Soul, and the Life. What is Death?

BARTHEZ, Lordat, and the Montpellier Medical School have created the doctrine of the *aggregate human*, which alone, we believe, can account for the true nature of man. This doctrine, which will serve us as a guide in the first part of this work, may be concisely stated as follows: —

There are three elements in man, —

1st, The body, a material substance;

2d, Life, or what Barthez calls the vital force;

3d, The soul, or what Lordat calls the inner sense.

The soul must not be confounded with the life; a mistake that many philosophers and scientific students have made. The life and the soul are essentially distinct. The life is perishable, while the soul is immortal; life is an ephemeral state, doomed to enfeeblement and destruction, while the soul is above every assault and escapes death. Like heat and electricity, life is a force engendered by certain causes: having begun, it comes to an end, and beyond this end it is nothing. The soul, on the contrary, has no end.

Man could be defined as a perfected soul resident in a living body.

This definition enables us to state what death is.

Death is the separation of the soul and the body. This separation happens when the body is no longer animated by life.

Plants and animals can live only in certain media: plants in the air or in the water, animals in the air, fishes in the water; and, when removed from their media, they perish immediately. Moreover, there are certain beings that live only in single media. Certain *vibrions* live only in azote gas, or carbonic acid gas. The germs of cryptogamic vegetables, which produce mould, develop only in watery infusions of vegetable matter. The fishes which live in the sea die when transferred to fresh or only slightly saline water.

Each living being has his special *habitat;* and the soul offers no exception to this rule. The medium, the *habitat* of the soul, is a living body. The soul leaves the body when the body ceases to live, as a man abandons a house when it is invaded and destroyed by flames.

Such is the doctrine of the triple alliance of the body, the soul, and the life, much the same as it was promulgated in the School of Montpellier; and such is, as a consequence of this doctrine, the mechanism of death.

We should add that this triple alliance of the body, the soul, and the life, is discoverable not in man alone: it is seen also in animals. Animals have a living body and a soul; only their souls are very inferior to ours in the number and compass of their faculties. With few needs, the animal has few faculties, which are all in a rudimentary condition. It is only in a much more complete development of the faculties of the soul that man differs from the inferior animals, to which he bears a resemblance, not limited to vital functions and anatomical structure.

It should be noted, in passing, that the School of Montpellier does not admit this last proposition touching animals. We shall have, however, to explain more at length, in another part of this work, our views on the differences which separate the man from the animal.*

CHAPTER II.

What becomes, after Death, of the Body, the Soul, and the Life?

AFTER death, in man and animals, the body — being no longer defended against destruction by the vital force — falls under the dominion of chemical forces. If the body of a dead animal, or a human body, be placed and kept in temperature below 0° + Centigrade, equivalent to 32° Fahrenheit; if it be shut up in a room entirely deprived of air; or if it be impregnated with antiseptic substances, it will remain unimpaired, —just what it was at the moment when life left it.

* Chapter XV.

This is the process of embalming. The various chemical substances with which the body is impregnated coagulate the albumen of the tissues, and preserve the animal substances from putrefaction. The same result follows if a body is kept between two layers of ice, or in a coffin enveloped on every side with ice, constantly renewed. Maintained in a temperature of 0°, the body will not undergo decomposition, because putrid fermentation cannot set in in so low a temperature.

In this manner have been preserved the entire carcasses of mammoths, elephants that belong to a species now extinct, and which lived in the quaternary epoch. In 1802, on the bank of the Lena, — a river which flows into the Frozen Sea, and traverses the country of the Yakoots, in those parts of Asia that lie about the North Pole, — there was found a perfectly preserved carcass of the gigantic pachyderm. The frozen earth, and the ice that covered the banks of the river where it was buried, had preserved it from putrefaction so thoroughly, that the flesh of the animal, which died more than a hundred thousand years before, served to regale the fishermen along the banks. In the northernmost regions, in order to preserve a human body unimpaired, it is only necessary to keep it constantly enveloped in ice.

Where the body of a man or an animal is exposed to the united influences of the air, the water, and a moderately high temperature, it undergoes a series of chemical decompositions, which terminate in its transformation into carbonic acid gas, ammonia, azote, water, and some compounds, gaseous or solid, which represent the later stages of destruction. Azote, carbonic acid, sulphuretted hydrogen, ammoniacal gas, and the vapor of water suffuse the atmosphere, or dissolve in the moisture of the earth. Later, these compounds, thus dissolved in the water that soaks the earth, are absorbed by the little roots of the plants that live in the soil, serving to nourish and develop them. As to the gases, they are at first disseminated through the air, and afterwards, falling back to the ground, dissolved in rains, minister also to the needs of vegetable life. Absorbed by the roots, the ammonia and carbonic acid, dissolved in the water which impregnates the soil, enter the ducts of plants and help to nourish them.

Thus the matter of the bodies of men and animals is not really destroyed: it simply changes its form, and in its new shape goes to the composition of new organic substances.

In this process, the human body only obeys the common laws of Nature. What happens to every organized substance, vegetable or animal, when exposed to the joint influences of the air, the water, and a moderately high temperature, happens to it also. A bit of cotton or woollen cloth, a grain of wheat, a fruit, ferment and dwindle into new products, just as the human body does. The sheet which covers the body is destroyed absolutely just as is the body itself.

But while the material substance which constitutes the body of man has only to transform itself, in crossing this globe, in order to pass from animals to plants, and from plants to animals, it is quite otherwise with his life. Life is a force. Like other forces — heat, light, and electricity — it has its origin and it transmits itself: it has a beginning and an end. Like heat, light, and electricity, — those physical agents which enable us to comprehend life, and which surely have the same essence and the same origin, — life has its productive and its destructive causes. It cannot relight itself when it is once extinguished; it cannot begin anew its course when its fated term has arrived. Life cannot perpetuate itself: it is a simple state of bodies, — a state fleeting, precarious, subject to a thousand influences and chances.

Life, then, is less important than the soul, which is indestructible and immortal. The soul is the essential element in all Nature. It has qualities active and positive wherever the other two elements, the body and the life, have negative ones. While the body becomes disintegrated and disappears, while life is annihilated, the soul can never vanish or become naught.

We have seen what becomes of the body and the life of a man after his death: let us inquire now what becomes of his soul.

The philosopher and the intelligent man, who know the immensity of the universe and the eternity of time, cannot but admit that our existence on earth is something definitive, — that human life is connected with something on this side of or beyond itself.

Man dies at thirty years of age — at twenty: he can live only some years — some months. The average length of life, according to the tables of Duvillard, is twenty-eight years. What is so short a period when compared with the duration of Time, with the age of the earth and the world? It is like a minute in eternity. Our life, then, so brief is it, can be only an accident; a fleeting phenomenon, hardly worth counting in the history of Nature.

In another view, the physical conditions of earthly life are truly detestable. Exposed to every kind of suffering, due as much to the defective organization of his body as to the external causes which incessantly threaten it; dreading extreme cold and extreme heat, feeble and wretched; coming into the world naked, and without any natural defence against the severities of climate, — man is indeed a martyr. If in a part of Europe and America the growth of civilization has secured comfort to the richer classes, what sufferings do not the poor of the same lands endure! Life is really a punishment to the majority of men who inhabit the insalubrious latitudes of Asia, Africa, and Oceanica. And before civilization came, in the periods in which lived primitive man, — periods so long that they mount up to a hundred thousand years before our day, — what then was the condition of the human race? It was an endless slavery of suffering, of dangers and pains.

The conditions of human existence are quite as bad when viewed from a moral as from a physical standpoint. It is a proved fact that happiness is impossible here below. When the Holy Scriptures tell us that the earth is a vale of tears, they only convey to us an incontestable truth under a poetic form. Yes: man is here on earth only to suffer. He suffers in his affections, in his unsatisfied desires, in the aspirations and soarings of his soul, continually driven back, bruised, broken by innumerable resisting obstacles. Happiness is a state that is forbidden to us. The few agreeable sensations that we transiently feel are paid for by the most cruel griefs. We have affections, only to lose and regret their dearest objects; we have fathers, mothers, children, only to see them some day die in our arms.

It is utterly impossible that a condition so anomalous should not be a definitive one. Since order, harmony, and tranquillity reign in the natural world, the same equilibrium must exist in the moral. If we see that all about us suffering is the common and constant rule, severe justice and violence dominating everywhere, force triumphant, victims trembling and dying under the hand of the oppressor, this state of things can be only temporary; only a moment of transition, an intermediate period which Providence has condemned us to traverse with rapid step to reach a better state.

But what is this new state? What is this second life, which is going to follow earthly existence? In other words, what becomes of the human soul after death has broken the ties which bound it to the body? This is the question that we are going to investigate.

We believe that after death the human soul passes into a new body, to be incarnated in another organism and constitute a being greatly superior to man in moral power, and ranking next above the human species in the hierarchy of Nature.

This being above man, in the scale of the living who people the universe, has no name in any language. Only the Angel, that the Christian religion honors and worships, can convey to us an idea of this being. Thus, Jean Reynaud called an *angel* a superior being, who, he admits, as heretofore, comes next beyond man, after death. Let us, however, waive this word, and give the name of superhuman being to that improved creature who, as we believe, is next higher to man in the ascending scale of beings in Nature.

CHAPTER III.

Where does the Superhuman Being dwell?

WE have just seen that, of the three elements which compose the *aggregate human*, there is one, the soul, which resists destruction. After the dissolution of the body, after life is extinct, the soul, freed from the material bonds which attached it to the earth, goes to feel, to love, to conceive, to enjoy freedom in a new body endowed with faculties more powerful than those which belong to humanity. It goes to constitute what we call the superhuman being. But where does this new being dwell?

All who have studied Nature know that life is lavished on our globe in truly prodigious proportions. We cannot take a step, we cannot cast a glance about us, without seeing everywhere myriads of living beings. The earth is a vast reservoir of life. Examine a blade of grass in a field, and you will find it covered with insects or inferior animals. But for such examination the eyes are not adequate: we must resort to the microscope. With the aid of a magnifying-glass, we discover that the single blade of grass is the home of a whole living population, which is born, dies, and multiplies with marvellous rapidity in this almost imperceptible domain.

From the blade of grass we can judge of all the vegetation that covers the globe.

The fresh waters that flow on the surface of the earth are likewise the repository of an immense number of organic existences. To say nothing of the plants and animals which live in rivers and streams, and which are visible to the naked eye, if you take a drop of water from a pool and place it on the object-holder of a microscope, you will find it filled with living creatures, which, though so small as to escape our sight, are not the less active in their places in the economy of Nature. We know how many inhabitants the sea conceals; but leaving out of the case the creatures plainly visible, — the fishes, the crustacea, the zoöphytes, as well as marine plants, — the creatures invisible to the naked eye, and which disclose themselves only under the microscope, so abound in the water of the sea, that a single drop, examined with the microscope, exhibits innumerable animals and microscopic plants.

From this drop of water we may judge of the whole

mass which fills the basin of the seas, and which constitutes three-quarters of the surface of our globe.

To convey an idea of the enormous number of living creatures that the seas hide and that they have hidden in former ages, let us cite here a fact well known to geologists, — that all our building-stone, all the limestone that forms mountains and chalk-banks, are wholly composed of the gathered fragments of the shells of mollusks, visible or microscopic, which filled the basin of the seas in the remotest time since the creation of the world. All soils are formed by the accumulation of shells. If life were distributed in the seas with such prodigality in the geologic periods, it must be distributed even to-day in nearly the same manner; for the actual conditions of Nature have not changed since the primitive days of our globe.

The air that surrounds us is, like the land and the sea, a vast receptacle of living creatures. We see but few animals in the air; but the philosopher, who goes beyond the mere appearance of things, can discover myriads of existences therein.

The air seems to us very pure and transparent, but it seems so because it is not sufficiently illuminated to enable us to perceive all the particles of foreign bodies which float in its depths. If you admit a ray — a pencil of solar light — into a close chamber, you see a bright luminous trail run through the room, which elsewhere remains dark. Every one knows that, owing to this powerful illumination, contrasted with the surrounding obscurity, the luminous trail is seen to be filled with light, tenuous, floating particles which flutter, rise and fall, in obedience to the disturbance of the air.

What appears in the atmosphere of a room so powerfully illuminated necessarily exists throughout the atmosphere which surrounds our globe; hence the air is everywhere filled with this animated dust.

What is this dust? It is composed almost entirely of living creatures, of the germs of microscopic plants (cryptogamic), or the eggs of inferior animals (zoöphytes). The so-called *spontaneous generation* of which much has been said lately, in France and elsewhere, is simply due to the organic germs which fill the air, and which, falling into the water or in the infusions of plants, give birth to mould, to growths which it has been attempted to refer to spontaneous generation; that is to say, to a germless creation, a causeless generation,—a gross error, for every living creature has parents, which a little knowledge and study can always discover.

The animals and plants called parasites constitute further evidence of the profusion with which life is distributed over the earth. Those animals or plants are called parasites, which live on other animals or plants, and subsist on the substance of their enforced entertainers. Every mammiferous animal has its parasites, —the flea, the bug, the plant-louse; man has the louse, the flea, the chigo. In like manner, every vegetable has its parasite. The oak gives a home and nourishment to lichens and divers cryptogamia; and even at its root may be found certain cryptogamia, such as the truffle. Here life is seen to implant itself, to ingraft itself on life.

More than this, parasites themselves have their special parasites, — smaller creatures, so small as to be invisible except with the aid of a microscope. Examine under this instrument a lichen on an oak; observe in the same way a flea, a plant-louse, and, thanks to the magnifying-glass, you will see the strange spectacle of a parasitic life superimposed on a parasite, and sustaining itself by the latter's substance. The alimental substance of a large vegetable passes to a visible, and from this to an invisible parasite. In this little space life is piled up and concentrated. Such a fact shows how lavishly it is scattered over our globe.

In our world, also, the fresh and salt water, and indeed the atmosphere, are inhabited by innumerable living creatures. Life superabounds on the earth, in the water, and in the air. Our globe is like an enormous vase, in which life has been accumulated, pressed down, and heaped up.

But earth, air, and water are not the only media of Nature. Beyond the atmosphere stretches another medium, well known to astronomers, and to which the name ether, or planetary ether, has been given. The atmosphere which surrounds our globe, and which goes with it on its journey through space, as in its rotation on its axis, is not very lofty, — only thirty to forty leagues, and diminishes in density according to its distance from the sun. At three or four leagues above the earth the air is so rarefied that it cannot be breathed by man and animals. For this reason, in aerial ascensions the limit of seven to eight kilometres * cannot be passed; for at this height the air is so thin and rarefied that it cannot be respired, nor can it counterbalance the pressure from the interior to the exterior of the human body. Beyond seven or eight kilometres, the density of the atmosphere diminishes rapidly, and at last the air fails absolutely. As we have said, at an elevation of thirty to forty leagues (120 to 160 kilometres) above the earth, the atmosphere terminates. At this point begins the fluid that astronomers and naturalists call ether.

This ether is a real fluid, — a gas analogous to the air which surrounds us, but infinitely more rarefied, and thinner. The existence of planetary ether cannot be doubted, since astronomers take its resistance into account in calculating the rapidity of motion of the heavenly bodies, as they take into account the resistance of the air in calculating the movements of bodies that traverse our atmosphere.

Ether, then, is a fluid which lies just beyond our atmosphere. It is distributed not only about the earth, but about the other planets. More than that, it is in all space: it fills the intervals which separate the planets. In fact, the planets, which with their satellites compose our solar world, move in ether. The comets also, in their immense journeys through space, circulate in ether.

---

* A kilometre is 1093.6389 yards.

The common people are disposed to believe that there is nothing beyond the air that surrounds the terrestrial globe, — that all is void. But there is no void in Nature. Space is always occupied by something: it is filled by earth, by water, by atmospheric air, or indeed by planetary ether.

We saw just now that life superabounds on our globe; that it swarms on the earth, in the air and the water. Is the ethereal fluid, which succeeds our atmosphere and fills space, also inhabited by living beings? Here is a question to which no scientific man has ever addressed himself. It would be strange, we think, that while life overflows, so to speak, in the water and the air, it should be absolutely wanting in the fluid which adjoins the air. All evidence goes to prove that the ether is inhabited.

But what are the creatures that live in the planetary ether? As we believe, they are superhuman beings, whom we regard as men brought into a new life and furnished with all kinds of moral improvements.

The chemical composition of planetary ether is unknown. Astronomical phenomena have taught us that there is such a fluid, but nothing is known of its components. We think we can say only that ether can contain no oxygen. Indeed, oxygen is the fundamental element of atmospheric air; and, as the respiration of man and animals becomes more difficult the higher they ascend, it must be concluded that it is the approach of an irrespirable gas which makes the upper parts of the air so unfavorable to us. He who, rising in a balloon, approaches the ether, is like a fish drawn half out of water into the air. The fish pants and palpitates on entering a medium which is fatal to it. It is so with a man who rises gradually from our atmosphere and draws near to the ether.

We may, it seems, conclude from this that there is no oxygen in planetary ether.

We should not wonder if planetary ether were composed of hydrogen gas excessively rarefied; that is, of a gas extremely light of its own nature, and infinitely subtilized by the absence of all pressure, which leads us to believe that it is hydrogen that constitutes the ether in which the planets move; that, according to observations lately made at the moment of the sun's total eclipse, a stratum of burning hydrogen gas surrounds the sun.

In all languages, the space above our atmosphere has received the same name: it is *heaven*. There then, in the place commonly termed Heaven, we fix the residence of superhuman beings. In this matter we concur with the popular belief and prejudices, and we gladly establish this agreement. These prejudices, these misgivings, in many cases epitomize the wisdom and observations of an infinite number of human generations. A tradition which uniformly prevails in all countries has the weight of a scientific demonstration.

Language and tradition agreeing, the most widespread modern religions — Christianity, Buddhism, and Mohammedanism — assign to Heaven the home of God's chosen people.

So science, tradition, and religions join hands in this matter; and the holy priest who, attending the royal martyr on the revolutionary scaffold, cried, "Son of the holy St. Louis, ascend to Heaven!" uttered a veritable scientific truth.

CHAPTER IV.

Do all Men alike, after Death, become Superhuman Beings? The Re-incarnation of the Wicked and of Infants.

DEATH, then, is not the end, but a change. We do not die: we are metamorphosed. The "situation" of death is not the *finale:* it is only one moving scene in the drama of human destiny. The agony is not the prelude to annihilation. It is only the enforced suffering which, in the order of Nature, attends every metamorphosis. Every one knows that the cold and motionless chrysalis is rent that the brilliant butterfly may emerge. Look at him at the moment of his escape from his temporary tomb. He is still panting and shivering with the pain of breaking the shackles that bound him. He must recover himself and regain his strength before he can dart in the airy domain which he is equipped to traverse. This is the image of our agony. In order to escape from the material envelope that we leave behind us, to raise ourselves to unknown spheres which await us beyond the tomb, we must suffer. We suffer in our bodies by physical pain, and in our souls by the anguish that we feel in facing our unknown future close at hand, which seems to us surrounded by terrible darkness.

But a difficulty presents itself here. Are all men alike going to become superhuman beings? In humanity there is an infinite scale of qualities and moral perversions. There is the good and the bad, there is the honest man and the criminal. Wherever we live, whatever our condition of mental culture, whether we are civilized or savage, enlightened or boorish; whether the question is of contemporaneous generations or of those who lived in remote ages, — everywhere, always, there is a universal morality, an absolute law of equity. Everywhere it is esteemed a bad action to kill a fellow-creature, to take another's goods, to maltreat children, to be ungrateful to one's parents, to quarrel with one's wife, to encroach on the liberty of another, to lie, to attempt suicide. From one end of the earth to the other, these actions have always been accounted bad. There are then; absolutely, even before Nature, good souls and bad. Must we believe that good and bad may alike be called upon to undergo that change of life which is going to bring us to the state of superhuman beings? Shall both classes be admitted to the joys of the new life which is reserved for us beyond the grave? Conscience, that exquisite sense that we carry within ourselves, and which never errs, — Conscience tells us that this cannot be so.

But how can the separation of the wheat from the tares be effected by the unaided forces of Nature? How can this choice be made, — a choice very difficult to explain, since the moral mingling with the physical seriously complicates a question of Nature? We can only give an individual opinion, not as a doctrine to be imposed upon anybody, but as simple evidence to be recorded.

It seems to us that the human soul, in order to rise into the realm of ether, must have attained an extreme degree of perfection, must have freed itself from all that weighed it down, must be subtile, light, purified: only at such a price can it quit the earth and soar toward heaven. Without intending to make a comparison, but simply to express our thought, we will say that the human soul seems to us a celestial aeronaut, who flies toward the sublime heights with speed and power, increased as his soul has been more refined and purified. Now the soul of a bad, a wicked, a vile, or a cowardly man, coarse and low, has not been improved, purified, and spiritualized: it is weighed down by bad passions, by the grossness of the appetites which it has not been able to destroy, or has even augmented. Unable to rise to the heavenly altitudes, this soul is forced to remain on our sad and miserable globe.

The bad and wicked man is not, then, we believe, called to enjoy, at least immediately, the blissful life which flows along in the serene regions of ether. His soul lingers here to begin a second life. And let us hasten to add, that he begins this second life without any recollection of his former existence.

It will be objected that to be born again, without any

recollection of the past life, is to fall into annihilation, to which materialists condemn us; that, indeed, it is identity that constitutes resurrection; and without recollection there is no identity: the individual, as an individual, would fall into annihilation if he were born again without memory.

This remark is just. If, after our resurrection to the state of superhuman beings, we had lost irreparably the recollection of our former life, we should be in fact the prey of annihilation. But it should be added that this loss of memory is of short duration. Forgetfulness of the past life is only a temporary condition imposed on our new existence: it is a kind of punishment. Recollection of the first life will return to the individual when, by the sufficient improvement of his soul, he shall have deserved to pass into the state of superhumans. Then he will recall the evil actions of his first earthly life, or of his several lives, — if he has had several probations to undergo, — and the thought of these actions will still be his punishment in the abode of bliss which he will have won at last.

To those who refuse their assent to these opinions, we will say that the question of punishments and rewards after death is the rock on which all the religions and philosophies have split. The explanation that we give of the punishment of the wicked is certainly preferable to the hell of Christianity, which is at once atrocious and absurd. The return to a second earthly life is a punishment less cruel, more reasonable and just, than condemnation to eternal torment. Here the punishment is proportioned to the sin. It is equitable and merciful, like the correction of a father. It is not an eternal penalty endured for a temporary fault: it is pitiful justice, which lays beside the punishment the means of avoiding it in the future. It does not forbid any return to good by an irrevocable doom for all eternity: it gives man a chance to tread again the path of happiness from which his passions have misled him, and to regain by his virtues the good things that he has lost.

So we think if, in its stay here below, the human soul, instead of improving, purifying, ennobling, and aggrandizing itself, has lost its strength and its natural qualities, — in other words, if it has been the property of a man wicked, coarse, and ignorant, base and vicious, — it will not quit the earth. After the death of this man, it will take up its residence in a new human body, losing the memory of its former existence. In this second incarnation, the unimproved and coarse soul, lacking all noble power and shorn of memory, will have to begin anew its moral education. The man who is born again a child recommences life with the uncultured and mean soul that was his at the moment of his death.

These reincarnations in a human body can be very numerous. They must be repeated until the faculties of the soul are sufficiently developed, or its instincts sufficiently improved, so that the man can raise himself above the common level of our species. Then, only, this soul, fittingly purified, freed from its imperfections, can quit the earth, and after the death of the body soar into space, and enter that new organism which succeeds that of man in the hierarchy of Nature.

We ought to add that infants dying at tender age, sucklings, or only a few months after birth, when their souls are yet undeveloped, have a similar fate. Their souls pass into the bodies of other children, and begin a new life.

We have only noted, in passing, this feature of our theory, which will be the subject of detailed investigation in another chapter.

CHAPTER V.

Does Man exist anywhere save on the Earth? The Planets and their Inhabitants.

IN the foregoing, we have reasoned as if the earth were the whole universe. Such nearly all men have thought to be the fact, from the organization of society down to the last century. We need a grand mathematical science, prolonged study, and optical instruments of the most improved kind, to correct the errors which result from our limited view of the earth and the heavens. Only great efforts of the mind, and a severe struggle against the evidence of our own senses, can convince us that the earth moves and the sun stands still. To recognize, in the uniform expanse of the heavens that meets our gaze at night, the place and the part of each one of these softly shining globes, demands patient and severe investigation transmitted and repeated from age to age: we need, in fine, an excellent scientific method. It is not strange that it has taken so long for man to penetrate the order of the universe, and that he has had in this matter, during thousands of years, only childish notions. The ancients — that is, the Greeks, the Romans, the Egyptians, the Orientals (except some truly wise men, who, by methods that we know not of, mastered the general mechanism of the universe, but concealed their knowledge from the common people) — knew of the universe only as the earth. Indeed, they had knowledge of only a small part of the earth, — Europe, Asia, and the north of Africa. The rest was a dead letter to the people of old. Following them and their example, the first Christians reduced the universe to the limits that they knew: they thought there was only one world, because they had seen only one. The earth to them was the universe. In the stars they saw only brilliant specks, and, as it were, silver studs that dotted the heavenly vault, to heighten its azure and charm the eyes of man in the silence of the nights. The moon was the natural lantern of the earth. In the heavens was a shining track that the sun traversed, and the torch of day was no greater than the lantern of night. The vast expanse which lay beyond the sun and the moon the ancients called the empyrean: for Christian and Mussulman, it was Paradise. It was at once the home of the clouds and of the light, the dwelling-place of the chosen of God, the blest and the good. Below, in the bowels of the earth, yawned huge abysses, gulfs, and caverns, gloomy abodes of the damned. It is known that Christ, after his death, descended into hell; after his resurrection, he returned to earth, and ascended in glory to Heaven, where God his Father awaited him.

This simple cosmogony, which only reported what the eye saw, has been accepted by all peoples in their infancy. Among the savage tribes of the world, — in America and Africa as in the ancient East, among the Romans as well as the Egyptians and the ancient Greeks, — this stupidly simple conception of the world, involving absolute ignorance of its constitution, has prevailed. On this basis, so thoroughly false, all the religions have been founded; and, we may add, they still rest there. The social customs of modern times are founded on the same errors. Language has consecrated them; for everywhere the earth is still called *the world*, as the ancients called it (*mundus*, *κόσμος*). Everywhere it is said that the sun moves from east to west, and that the stars rise or set. Poetry has placed its eternal seal on the vicious system, and has also — to repeat a word — consecrated it with the spell of imagination and genius.

Modern astronomy has shattered the lying heavens of antiquity: it has dissipated the so-called heavenly vault sown with brilliant specks, and replaced it with a mere mass of colored air. It has disclosed the true office of each one of the heavenly bodies that are visible to us by day or by night. It has established infallibly the place of the earth in the universe, and, it must be acknowledged, this place is singularly insignificant.

We know now that the earth, far from being itself the world, is only an imperceptible point in the world. Compare it only with the sun: we know that our globe is thirteen hundred thousand times smaller than that luminary. See how far we are beyond the idea of the ancient Greeks, who believed that the sun was as big as Peloponnesus.

The earth has been, moreover, ousted from its high estate. Once it was thought to be unique and unrivalled: now it is ascertained that there is an infinity of other globes like it, and that it is therefore nothing but an individual in a group of individuals that resemble it. It is known that the earth ranks among the planets, that it is but a planet of our solar system.

What, then, is a planet? asks the reader. A careful scrutiny of the heavenly bodies at night will make him understand. Let him examine, some fine night, the luminary that an intelligent observer will point out as Mars or Jupiter, and let him assure himself, at a fixed hour, of its position. Then, some hours later, let him look again for Mars or Jupiter, and he will find that the planet has changed its place relatively to the other bodies. Let him do better than this: let him look at Jupiter through the telescope of an observatory, or one of those open-air astronomical instruments to be found in public places in Paris and other great cities. He will then see Jupiter or Mars move before his eyes; while the other heavenly bodies in the field of the glass remain fixed, Jupiter or Mars will pass beyond it.

There are, then, transient and fixed bodies in the heavens. The transient ones are planets (*planetes*, from πλάνος, wandering): the fixed ones are the stars.

It is easy to distinguish the planets from the fixed stars. The stars shed a twinkling light (whence comes their name, derived from the Latin *stellare*, to glitter): their radiance is tremulous. On the other hand, the light of the planets is always quiet and steady.

This is because the stars shed their own light. As will be stated farther on, the stars are as much suns as is our sun. They illuminate worlds like our own, that only their great distance prevents us from seeing. Planets do not shine of themselves: they only reflect, like gigantic mirrors, the light of the sun which falls upon and renders them visible to us.

Thus planets are stars that move. They revolve around the sun. The Earth, being a planet, is a star that moves, that revolves around the Sun.

But the Earth is not the only planet in our solar system. There are seven others, that do not differ essentially from it. Here are the names of the eight planets that compose our solar system, arranged according to their distance from the Sun: Mercury, Venus, Earth, Mars, Jupiter, Saturn, Uranus, and Neptune. It should be added that between Mars and Jupiter there is a mass of small bodies which appear to be fragments of broken planets: they are called Asteroids. They number more than a hundred: it is easy to imagine them united in a single group which would constitute the ninth planet.

So the Earth enjoys no special distinction. Her part in the universe is equally shared by other heavenly bodies, and there is no ground for the pre-eminence that the ancients accorded her. She is only the third of the eight planets of our system. She is not the largest; for Jupiter is 1400 times, Saturn 734 times, Uranus 82 times, Neptune 105 times larger than she. She is not the smallest; for Mercury and Mars are smaller, and Venus is of about the same size. The Earth is not the warmest of the planets: this is evident from her distance from the Sun. Mercury and Venus are warmer: Mars and the planets beyond him are colder.

The Earth, we repeat, simply makes a part of a class or a group of stars; and she is only an individual of this class or group.

These considerations lead us to a very important deduction. Since there is nothing to distinguish the Earth from the other planets of our solar system, we must find in the others what we find here: there must be in these planetary globes air and water, a hard soil, rivers and seas, mountains and valleys. There must be found also in them vegetation and trees, and tracts covered with verdure and shade. There must be in them animals and even men, or at least beings superior to animals, and corresponding to our human type.

But is this possible? Is it true? Are the planets that, like the Earth, revolve around the Sun, and in the same time with the Earth, — are they of the same physical constitution as the Earth? are they covered with vegetation? are they inhabited by animals and men? These are the questions that we have now to investigate.

To solve this grand and difficult problem, we will simplify it. Instead of surveying the eight planets of our solar system, we will, to begin with, look at only half of them. In other words, we will divide them into two groups, according to their distance from the Sun.

The four planets nearest the Sun are Mercury, Venus, the Earth, and Mars. We will first try to describe this group, and then will inquire if living things, like the plants and animals of our globe, can exist in Mercury, in Venus, and in Mars. After this, we will survey from the same point of view the planets farthest from the Sun; that is, Jupiter, Saturn, Uranus, and Neptune.

It is plain that Mercury, Venus, the Earth, and Mars form what we may call a natural group. They are of the middling size, not very distant from the Sun, and differ little in temperature. They are not unlike in the duration of their years and in the length of their days, and in size.* What is true of one of them, then, must be true of the others. On the other hand, Jupiter, Saturn, Uranus, and Neptune, planets of considerable magnitude, all rotate on their axes with great rapidity, which renders their days extremely short. They all have numerous satellites. These considerations suffice to justify the division of the planets into two groups, that

* This resemblance is not very close, as Mercury has only 88 days, and, as the author says, p. 38, it "is considerably smaller than our globe." — Tr.

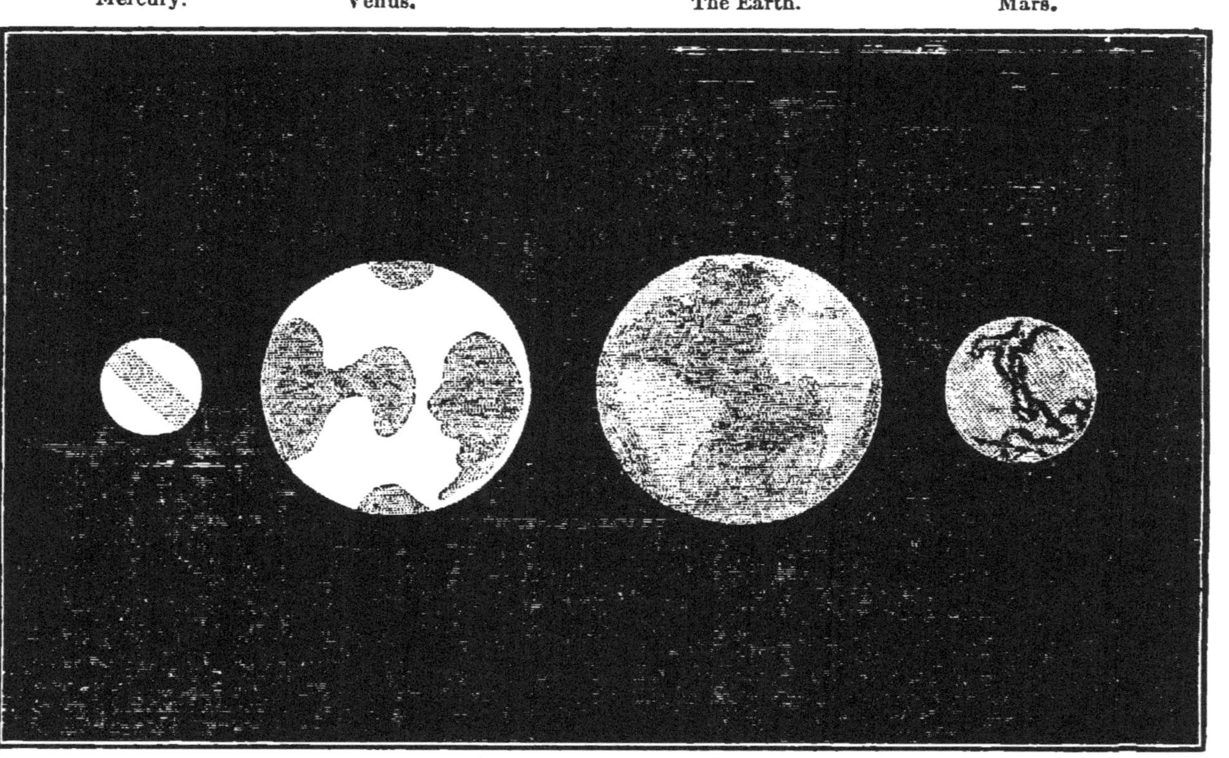

Fig. 1.—Comparative magnitudes of the planets Mercury, Venus, the Earth, and Mars.

we have made in order to help our readers to comprehend the facts that we are going to survey; that is, in studying the question of the habitability of the planets.

CHAPTER VI.

Description and Geography of the Planets Mercury, Mars, and Venus. Are these Planets inhabited?

MERCURY is the planet nearest the Sun, being distant from it only 14,000,000 leagues, which is near neighborhood in astronomy. Therefore the Sun, seen from Mercury, must seem to be six or seven times larger than it seems to us: it must look like an enormous disk all glowing with fire. Six or seven times as much heat and light fall upon Mercury as upon our globe.

Mercury revolves on its axis just as rapidly as the Earth. Our days are 24 hours long, and those of Mercury 24 hours, 3 minutes. Being nearer the Sun, Mercury makes its journey around that luminary quicker than does the Earth: its year, instead of 365 days like ours, has only 88 days.

Mercury's year being shorter than the Earth's, its seasons are shorter than ours: they last only 22 days, while ours are 90 days long. Its seasons are also more diverse, because the changes from heat to cold, or to a mean temperature, are more abrupt in Mercury than in our globe.

It is known that the only cause of the inequality of

the seasons, as well as of the days and nights, in the planets, lies in the inclination of each on its axis. If they revolved around the sun on vertical axes, joining their north and south poles, the distribution of solar light and heat would be perfectly equal in the same latitudes; along each parallel there would be perfectly equal and regular illumination and heating of the planet; the differences in heat and cold would depend only on their greater or less distance from the Sun. But this verticality exists in only two or three of the planets of our system. The others — and among them Mercury, Venus, the Earth, and Mars — have a perceptible inclination on their axis. They revolve at an inclination, — "out of plumb," so to speak, as if they had received a tremendous blow on the shoulder which forced them out of their original regular attitude. Hence results a great difference in the duration of light, and consequently of the heat that the inclined planets receive from the horizontal rays of the Sun. From this cause come the inequality in the length of days and nights, and the diversity of the four seasons in the same parallel.

The inclination of the Earth on its axis is 23°; a considerable deviation, and which occasions great differences in the length of days and seasons in the several parts of our globe. In Mercury the inclination is prodigious, not less than 70°. This planet leans on itself, as if it were going to fall. From this peculiarity results a great variableness of heat and light on the same parallel, and seasons whose abrupt changes it must be difficult for the inhabitants of the planet to endure.

Mercury, like the Earth, is enveloped in an atmosphere of considerable density.

Observations of this planet with the best telescopes have shown that its surface bristles with mountains much higher than those of the Earth, although Mercury is considerably smaller than our globe.

NOTE. — Milton says, in "Paradise Lost," that before our first parents sinned perpetual spring reigned on earth : but that, as soon as Adam and Eve had eaten the forbidden fruit, angels with flaming swords were despatched from heaven, to bend the poles of the earth more than 20°. It is lucky for us that the angels did not make the inclination still greater: if they had, the seasons would have been even more abrupt and faulty.

It is well known that Fourrier declared it to be possible for man to exert a power sufficient to readjust the earth on its axis, and to restore the equality of the seasons and perpetual spring. He forgot only one point, — to indicate the mechanical process by which this power could be exercised. Thus a drowning man thought to save himself by seizing his own hair while struggling in the water.

Mercury is, in fact, five times smaller than the Earth. This is shown in Figure 2, in which T represents the Earth, and M Mercury. Its diameter is only 1,200 leagues in round numbers, while that of the Earth is 3,000.

The heavenly bodies seen from Mercury would seem just such as they appear to us, and at the same points at which we perceive them, according as, borne through space, we are brought into the presence of the several stars of the universe.

The same spectacle is given, indeed, to all the planets of our solar system: the heavenly vault is the same to them. In fact, all borne along together, sharing in a common movement, they must have in their progress the same perspective. If there are dwellers in the planets, they must see the vault of heaven just as we see it from the Earth.

Fig. 2.—Comparative magnitudes of the Earth and Mercury.

But if the stars seem to the inhabitants of the planets just as they seem to us, it is otherwise with the planets themselves.

To the dwellers in Mercury, for instance, the other planets cannot offer the same aspect that they present to us. Venus must seem to them larger and more powerfully illuminated than we find it, brilliant as it is; in fine, Mercury is nearer to Venus than to the Earth. As Venus shines six times more brightly seen from Mercury than when seen from the Earth, Huygens believed that Venus must serve as a moon to Mercury, lighting its nights.

Seen from Mercury, Venus and the Earth undergo continual variations of illumination, owing to their changes of place in relation to the Sun: they have *phases*, the astronomers say, such as the moon presents to us.

We give an idea, in Figure 3, of the manner in which Mercury is viewed, according to the place it occupies opposite the Sun; in other words, of the *phases* of Mercury.

Fig. 3.—Phases of Mercury.

It is probable that the dwellers in Mercury do not see all the planets of our system, owing to the extreme distance of the latter. They see, perhaps, the great Jupiter and even Saturn; but it is not likely that they can see Uranus. As to the invisible Neptune, who conceals himself beyond the limits of our world, he is certainly unknown to them.

Next to Mercury, in respect to distance from the Sun, comes Venus, that bright forerunner of our nights, the star of twilight, which has always excited by its brilliancy the sympathetic aspirations of thoughtful souls.

At a distance of 27,000,000 miles from the Sun, Venus receives twice as much light and heat as does our globe. Her days are hardly so long as ours,—

23 hours, 21 minutes. Venus's year is, of course, shorter than the Earth's, on account of her closer proximity to the Sun: it is only 224 days long. Her seasons last only two months each. Her size is nearly the same as the Earth's.

The proportionate magnitudes of these two planets are exactly represented in Figure 4, in which T stands for the Earth, and V for Venus.

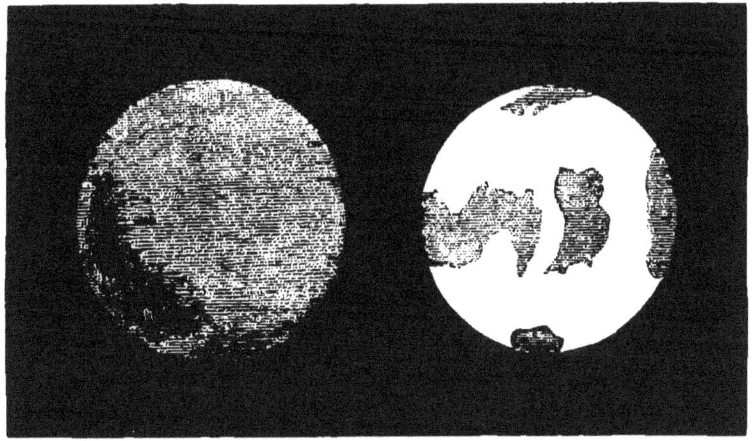

Fig. 4.—Comparative magnitudes of the Earth and Venus.

Venus is surrounded by an atmosphere of about the same height as the Earth's, and on the verge of her globe may be witnessed the phenomena of twilight and dawn, which are due to the atmosphere. Clouds almost constantly envelop Venus, and pour upon her rains which must form rivers and seas. These waters must refresh the plains parched by the intense heat of the burning sun. Mountains tall and slender, visible through a good telescope, ridge her continent. These, no doubt, serve as distributing reservoirs of the waters that pour along their slopes. Some of them have been approximately measured by means of photographic views taken in the telescope. They are estimated to be 40,000 metres (about 125,000 feet) high; an enormous height, when it is remembered that the highest mountain on our globe (the Kaurisankar in the Himalayas) is only 10,000 metres.

The seasons of Venus are even more abrupt and dissimilar than those of Mercury. Her axis is, in fact, inclined 75°. M. Babinet describes as follows the effect of this great inclination on the seasons of this planet:—

"The planet which certainly presents the most remarkable climatologic peculiarities is Venus, which in bulk, and distance from the Sun, is almost the exact counterpart of the Earth. She turns very obliquely on herself. If we take the Earth for a point of comparison, the Sun in summer comes almost above Sienna in Egypt, or Cuba in America. The obliquity of Venus is so great that in summer the Sun attains latitudes higher than those of Belgium or Holland. It follows from this that the two poles, subjected in turn to an almost vertical sun, which never sets (and this at intervals of four months, since the year of this planet is only eight months long), do not permit snow and ice to accumulate. This planet has no temperate zone: the torrid and icy zones encroach the one upon the other, and rule successively over the regions which in our world constitute the temperate zone. Hence results constant agitation of the atmosphere, agreeably to what observation has taught us as to the difficulty of seeing the continents of Venus across the veil of her atmosphere, incessantly disturbed by the rapid variations of the height of the Sun, of the duration of days, and the transports of air and moisture that render the rays of the Sun twice as powerful as those that come upon the Earth."

The winter in Venus, therefore, is as cold as in the Earth, and the summer is infinitely hotter. Furious winds incessantly sweep the atmosphere, owing to the considerable variations of temperature at opposite points of the globe.

Except in this respect, Venus must very strongly resemble the Earth. Like the Earth, she enjoys all the grandeur of the light. Clouds traverse the heights of her atmosphere, rains wash her surface. On the horizon, at morning, the dawn colors the east with its silvery rays, forerunning the huge disk of the Sun: at evening this disk sinks in billows of purple and gold. The telescope, in fine, enables us to see dawn and twilight on the surface of Venus.

From this planet the heavens seem studded with stars, just as we see it, according to the general remark that we made about Mercury.

Seen from Venus, the Earth must appear exceedingly luminous, — brighter than Venus herself appears to us, for she can be seen close at hand when illuminated by the sun. Perhaps the Earth serves to light by reflection the nights of Venus, as the moon lights the Earth by reflecting the solar rays.

Our moon also must appear most brilliant when seen from Venus.

From Venus may be seen, almost as we see them, the planets Mars, Mercury, Jupiter, and Saturn. Uranus is probably invisible by reason of his great distance; and Neptune, for a still stronger reason of the same nature.

Leaving Venus, we come to the Earth, which is the next planet in the order of distance from the Sun.

Distant from the Sun 38,000,000 leagues, the Earth is of about the same size with Venus. Her diameter is in round numbers 3,000 leagues (exactly 12,732,814 metres). Surrounded by an atmosphere, she revolves on her axis in 24 hours (23h. 56m. 4s.), and in 365 days, 5 hours, accomplishes her journey around the Sun.

The inclination of the Earth's axis to the ecliptic is 23°, which occasions, as we have seen, the difference of days and nights, the inequality of the seasons, according to the latitudes.

The Earth is favored, unlike Mercury, Venus, and Mars, with a secondary star called her satellite: it is the moon. Only 96,000 leagues from the Earth, the moon makes the journey about her in 27 days.

It would transcend the scope of this work to give any description whatever of our globe. We must assume that it is sufficiently known to our readers, and pass on to the planet next in the order of distance from the Sun.

Next to the Earth, going from the Sun, is Mars. All hypotheses are possible, are they not, reader? Suppose, then, that one of the many balloons that Paris sent to the Departments during the Prussian siege in 1870 had risen to a height of about 15,000,000 leagues from the Earth in the direction of the Sun, and that the aeronaut who managed this hypothetical balloon still lived, what would happen? It would happen that, at this distance of 15,000,000 leagues from the Earth, the balloon, yielding to the gravity which would draw it toward Mars, would fall upon that planet. Now — and this is what we are driving at — the balloonist, setting his foot on

Mars, would think himself to be on the Earth, and could not undeceive himself. Soil, mountains and hills, rivers and seas, vegetation, — he would find all this in the planet his steps were treading; so that it would be impossible for him to know if he had fallen upon an unfrequented corner of the earth, like Australia or Polynesia, or upon a planet unknown to us.

We have made use of this illustration the better to show the resemblance, which is truly remarkable, that exists between the Earth and Mars. Physical, geographical, climatological conditions; nights and days; seasons, celestial vistas, — in all these points the two planets are alike, with the single difference that Mars is smaller by half than the Earth. This correspondence is illustrated in Figure 5, in which T represents the Earth, and M Mars.

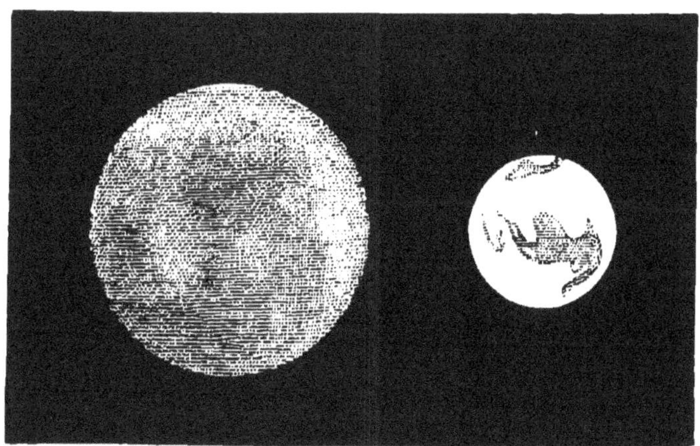

Fig. 5. — Comparative magnitudes of the Earth and Mars.

Here, moreover, are the exact figures of the heavenly geodesy of Mars. Distant from the Earth 20,000,000 leagues, and from the Sun 58,000,000, Mars has a diameter of 1,650 leagues, that of the Earth being, as we have said, 3,000 in round numbers. Its days are 24 hours, 39 minutes in length, and its year is longer than that of the Earth, because it is farther than the Earth from the orb of day. The year of Mars is 22 months and 11 days. The inclination of its axis differing but little from that of the Earth, the seasons, like the hours of day, are distributed much like those of our world. Only, being farther from the Sun than the Earth, it is less heated and illuminated than the latter: it receives only about half as much light and heat as we do.

Mars is surrounded by an atmosphere, which — thanks to the effects that it produces — can easily be distinguished through a telescope. If we observe it at the moment of its closest proximity to us, — that is, about 14,000,000 leagues distant, — we see at the two poles two spots of dazzling white: these are snow or polar ice, like those at the poles of the Earth. Looking between the two poles, and drawing near the equator, we see continents and seas. Through the telescope this planet appears like a great geographical terrestrial globe, looked at from a considerable distance.

Figure 6 represents the aspect of Mars when it is nearest the Earth. At the north and south poles may be seen the two white spots made by snow.

The continents of Mars, seen through the telescope, exhibit reddish tints like the ochreous soil of the Earth. Some astronomers have attributed this peculiar hue to vegetation which would prefer this tinge. Since we have plants with red leaves, or which become red in

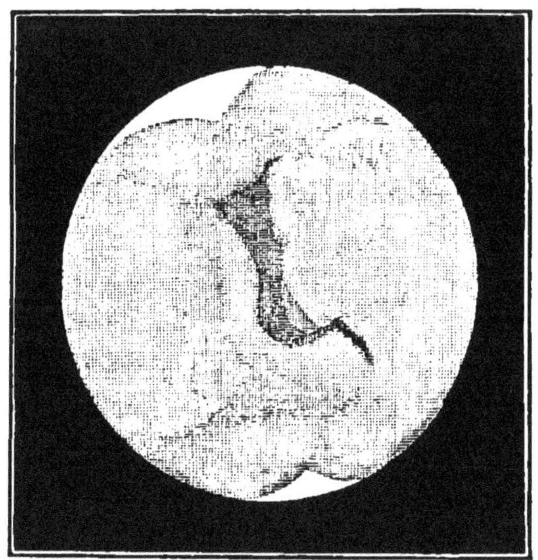

Fig. 6. — The planet Mars as seen through the telescope.

autumn, it is not impossible that vegetation in Mars takes on this particular color. In this way we may account for the rosy hue that this planet sometimes wears to the naked eye.

We have just said that the inclination of the axis being nearly the same in Mars and the Earth, the seasons have the same course in both. If you watch with the telescope the growth or diminution of the snowy patches which are visible at the poles of this planet, you will understand that the ice melts or accumulates at the poles, at the same period as on the Earth. On the latter the ice of the northern hemisphere melts or diminishes during the summer, and that of the southern during the winter. In like manner the snow patches on the northern hemisphere of Mars are seen to diminish in summer, and to melt toward 60° of latitude, as they melt with us at about 70°. In the winter the snow patches of the southern hemisphere are seen to diminish, as on our globe the snows of the corresponding hemisphere melt away in the same season.

Thus the climatology and its phenomena, — which depend upon the physical character of the globe, — cold and heat, rain or drought, the length of days and seasons, specific gravity, and quality of soil, are the same in the Earth and in Mars. It is not paradoxical to claim that an inhabitant of the Earth, transported to Mars, would find it difficult to distinguish that planet from the Earth, his home.

We have now taken note of the astronomical data, the physical constitution, and climatology of the four planets nearest the Sun, — Mercury, Venus, Mars, and the Earth. Let us proceed to draw the conclusions which legitimately result, we think, from the harmony of facts.

Mercury, Venus, and Mars, like the Earth, have days and nights; and the length of days differs little in all. Mercury, Venus, and Mars have a hard soil, and a surrounding atmosphere like that which envelops the Earth. They are often, like our globe, covered with clouds which produce refreshing rains. Water exists on these four planets in vaporous, solid, and liquid states. The temperature varies between Mercury and Mars, but in small proportions. Twice as much solar heat falls on Venus as on the Earth, and one-half as much on Mars. Mercury is the hottest of these four globes.

A moderate temperature, air and water, hatch and nourish organic life, — the life of plants and animals. Plants and animals live on the Earth because they find here these requisites. All these conditions are found in the three other planets that we are surveying. Mercury, Venus, and Mars have each air, water, and a moderate degree of heat. May we not conclude from this fact that in these planets, as in the Earth, there are organized existences, plants and animals?

We do not hesitate to draw this conclusion. We do not hesitate to declare that the planets, being habitable, are inhabited; that the demonstration of this fact of habitability demands and logically involves the fact of habitation. Otherwise, the Creator would be illogical; his acts would be inconsequent; he would aim at an end without attaining it.

But reasoning, induction, and analogy are not the only paths that lead us to people the planetary globes with organized existences. We have direct proofs of this fact. We can show and touch natural substances that have come from the planets.

Aerolites, now called meteorites, and which used to be called stones fallen from the heavens, are really, as this common name expresses it, stones hurled from planetary spaces by some accident of the Universe. For a long time the Paris Academy of Sciences scouted this idea; and, yielding to academic opinion, naturalists refused to believe in stones fallen from heaven. But to-day there is no doubt of it; and philosophers are agreed in the opinion that aerolites are fragments of planetary bodies which, cast into space and coming within the sphere of our globe's attraction, fall upon its surface. From what heavenly bodies can they come but from those which are nearest to us; that is, Mars and Venus, between which the Earth is placed? Chemical examination of these aerolites discloses to us the constituents of the soil of Mars and Venus, our neighbors.

There are now five collections of aerolites in Europe and America. Probably these stones do not now fall so frequently as formerly in America, when the savage inhabitants of that country drew supplies of iron from them for making knives and domestic utensils. Yet enough are seen to fall in our day to make up valuable collections. The Museum of Vienna contains one hundred and sixty-two fine specimens; the British Museum has one hundred and sixty; and the Jardin des Plantes, at Paris, an equal number. One of the aerolites in the British Museum weighs not less than 634 kilogrammes.* It is plain that chemical analysis of such bodies must tell us the composition of the soil of those planets which are adjacent to the Earth.

Such analyses have often been made. In Germany, Reichenbach has made a great number; in France, MM. Daubrée and Stanislas Meunier have published the results of their researches in the same direction. It has been ascertained that the mineral substances that exist in these aerolites are the same that compose our soil; that is, oxide of iron, oxides of nickel, of cobalt, and of manganese, magnesia, lime, silica, copper,

* About 2,200 pounds.

and sulphur. Some peculiar metals, unknown on our globe, have been discovered in these aerolites.

Let us not forget to add that water has been found in them. Indeed, the oxide of iron is often present in aerolites in the form of hydrate; that is, combined with water. Thus chemical analysis proves that there is water in the planets.

Another important revelation effected by chemical analysis relates to carbon. Some aerolites contain graphites mixed with oxide of iron or coal. There is coal, then, in the planets beyond the Earth. But coal is a fundamental element of animals and plants; and if there is coal in Mars and Venus, we must conclude that there is organized life there.

This life is not, moreover, the simple creature of reasoning and analogy: it has been seen and touched. An aerolite, which fell at Orgueil in 1864, was analyzed at Paris by M. Daubrée, who found in it, mixed with oxide of iron and water, a genuine organized substance, — peat. Peat is simply the product of the gradual decomposition of vegetables under the action of water. There were vegetables, then, in the planet from which came the Orgueil aerolite. By consequence, there are vegetables in the planets which neighbor our own.

The "Scientific Press," in giving an account of Reichenbach's studies of aerolites, says: "These fragments contain not only metals and ordinary metalloids, but also coal; that is, a body whose origin we can always trace to organized life, and which, if it is possible to apply to unfathomed regions truths that we see about us, must have been animalized."

Already, however, the existence of organic matter in aerolites had been suspected. Organic vegetable matter was found in one which fell at Sienna, May 16, 1830, and which was analyzed by Professor Giuli. A French naturalist, M. Angelot, in a paper read before the Geological Society of France, Feb. 17, 1840, and published in the Society's "Bulletin," remarked that there were found in stones fallen from the heavens "oxygen, carbon, hydrogen, as well as water combined as a hydrate with oxide of iron, almost the only form in which it could come to us." This observer adds that "we have proof that there are beyond our globe chemical elements of a vegetable kingdom like our own."

Modern studies of aerolites have special importance in the question that we are now discussing. Aerolites seem like messengers sent to us, across space, by the extra-terrestrial worlds, to tell us of the composition of their soil. And chemistry shows us that animal and vegetable kingdoms must exist in those planets, as on the Earth.

But, it will be objected, it is so hot in Mercury and Venus, and so cold in Mars, that vegetables and animals cannot live. This objection has been frequently urged. In Mercury, it is said, herbage would be burned, the harvest dried up by the sun, animals asphyxiated by the heat, man blinded and crazed by the excess of light. It is true that Mercury receives from the Sun six or seven times more light than the Earth, and that

Venus receives twice as much. But it is not beyond question that all the heat emitted by the Sun reaches the surface of Mercury or Venus. A very dense atmosphere, as we have said, surrounds these globes. If this atmosphere has a great power of absorbing heat, a conducting power which disseminates rapidly throughout its whole mass the heat which reaches it at a single point; if the soil of the planet is by its physical character enabled powerfully to reflect light and heat, while absorbing but little, it is possible that the atmosphere retains and distributes in itself two-thirds of the heat emitted upon it by the Sun. As the composition of the atmosphere of the planets is unknown to us, we can affirm nothing as to its properties; but we may believe that the physical constitution of this atmosphere can mitigate in a great degree the heat of the Sun, as it does so wonderfully on the Earth. An atmosphereless globe, like the moon, is chilled and heated in extreme degrees: it has nothing to counterbalance the effects of the source of heat or of the cause of cold. But a globe wrapped, like the planets, in a gaseous envelope, is protected by it from the excess of heat or cold. We believe that the solar heat transmitted to Mercury may be weakened to the extent of two-thirds by atmospheric absorption, and that, so far from being six times as hot as the Earth, that planet is only twice as hot. The mean temperature of a place like Paris being $+10°$,* it must follow that the mean temperature of a place in Mercury, in the same latitude and longitude as Paris,

---

* Réaumur.

would be + 20° or + 25°, which is not extreme, and represents a thermal condition most suitable to us.

The same reasoning may be applied to Venus, to prove that in that planet, which receives twice as much heat as the Earth, the temperature may be just about the same as on our globe.

On the other hand, it may be proved that it is not so cold in Mars as to render the life of organized beings difficult or impossible. A certain physical constitution of the atmosphere in that planet can weaken the moderate degree of cold which is due to its distance from the Sun. Perhaps, too, its own heat supplies the deficiency of solar heat.

Thus the argument against the possibility of the existence of organized life in the three planets that we are surveying — an argument derived from their distance from the Sun — has, for the scientific investigator, no substantial basis.

It must be understood, however, that in contending that Mercury, Venus, and Mars must contain living existences, vegetable and animal, we do not mean to say that these vegetables and animals are like those of the Earth. Astronomers who have combated the theory of the habitability of the planets, and even writers who sustain it, have fallen into an error of reasoning. They have copied the types of dwellers on the planets from the type of dwellers on the Earth. They have given to the former the figures and qualities of the latter. We must reason in another way, and must not transport to Mercury, Mars, or Venus creatures made for a terrestrial medium. It is very evident that our plants and animals would die in Mercury, Mars, or Venus, — media which are not adapted to their organs. But this is not the question. The question is, Are the physical and climatologic conditions of Mercury, Venus, and Mars favorable to life? If they are, — if air, water, and temperature permit organized beings to grow and multiply in those planets, — we may trust to Nature to create figures and organic types in harmony with the media that they are destined to inhabit.

Organized existences cannot grow except where their life is guaranteed by the composition of the media in which they are; and we believe that the media of Mercury, Venus, and Mars are perfectly adapted to such life. As to the forms which these existences can take, it is impossible for us to imagine them. We can only say that they must be consistent with the conditions of the planet, and that the vegetables or animals of Mercury, Venus, or Mars, transferred to Earth, would perish, as our men and animals would perish in the other planets.

Many astronomers, moreover, have taught the doctrine of the habitability of the planets, to say nothing of modern *savants*, who, almost without exception, have accepted it as a fact.

After the beginning of the new astronomy, the celebrated Huygens, following in the steps of Fontenelle, wrote a remarkable work, the "Cosmotheoros," in which he demonstrated that there were living creatures, and even men like us, in the planets. Speaking of the inhabitants of Mercury, Huygens speculates as to the degree of scientific knowledge possessed by planetary men. He is convinced that they are expert astronomers, the close neighborhood of the Sun enabling them to obtain exact information as to the course of the planets around that luminary. He goes so far as to inquire about the instruments employed in astronomical observations by the inhabitants of Mercury, and debates whether wood, copper, and glass enter into their composition.

Another astronomer, Christian Wolf, who lived at the beginning of the last century, went still farther. He undertook to determine the size of the people of Jupiter, and estimated it at fifteen feet. Here are the curious calculations by which he reached his estimate.

He remembered, first, that light dilates the pupil of the eye in man and animals: this dilatation of the pupil must increase the dimensions of the globe of the eye. He concluded that this expansion of the globe of the eye must produce an enlargement of the whole figure of the animal. From this a knowledge of the degree of intensity of light on one planet would enable him to estimate the size of the inhabitants of Jupiter. Now we know the intensity of solar light on the surface of Jupiter by the distance of that planet from the Sun.

But we will quote this curious passage from the German astronomer:—

"It is taught in optics that the retina [pupil] of the eye is dilated by a feeble, and contracted by a strong light. The light of the Sun being far feebler to the inhabitants of Jupiter than to us, by reason of their greater distance from its source, it follows that the pupils of their eyes are much larger and more dilated than ours. Now we observe that the pupil is always in proportion to the globe of the eye, and the eye in proportion to the rest of the body; so that the more developed the pupil in an animal, the greater is his eye and his whole body. To determine the size of the people of Jupiter, we must consider that the distance of Jupiter from the Sun is to the distance of the Earth as 26 is to 5; and that, by consequence, the light of the Sun in relation to Jupiter is to that light in relation to the Earth in duple proportion of 5 to 26. Moreover, experience teaches us that the dilatation of the pupil is always more than proportional to the increase of intensity of the light; otherwise, one body placed at a great distance would seem as clearly defined as another placed nearer. The diameter of the pupil of the people of Jupiter is, then, to the diameter of ours in greater proportion than 5 to 26. Suppose this proportion to be 10 to 26, or 5 to 13. The average height of human beings being about five feet, four inches, we must conclude that the average height of the inhabitants of Jupiter is fifteen feet."

These calculations are very ingenious; but unfortunately the leading idea lacks a basis, for it is not accurate to say that the size of the globe of the eye is in proportion to the size of the animal. The elephant and the whale, Colossi of the animal kingdom, have very small eyes. The bee, on the other hand, has a very large eye. Birds of prey have large eyes, and are yet not of very great size. The skates that Biot, during his expedition with Arago to the Island of Formentera in Spain, found on the bottom of the sea, very deep and consequently very dark, had eyes of inordinate size protected by large and hard bones; yet the size of these fish was not great.

Although founded on an erroneous assumption, these calculations of Wolf deserve to be reported here by reason of their remarkable conclusion.

Fig. 7.—Size of the planets Jupiter, Saturn, Uranus, and Neptune, compared with the Earth.

CHAPTER VII.

Description and Geography of Jupiter, Saturn, Uranus, and Neptune. Are these Planets inhabited? General Conclusion as to the Existence of Living Beings in all the Planets of our Solar System.

CONTINUING our journey in the heavens away from the Sun, we first reach, after leaving Mars, the ring of Asteroids. We will not pause at this gathering of little stars, which, without doubt, is only a collection of severed fragments, more than a hundred in number, of some planet which once existed at this point, and was broken in pieces by some mighty convulsion of the Universe. Names have been given to these little stars, as well as to the important planets,—Vesta, Pallas, Circe, &c. Maximilian and Feronia lie at the two extremities, regarding them as to their distance from the Sun.

These remains of a broken star continue to revolve around the Sun like the planet which they originally composed.

Next to the Asteroids comes the great Jupiter, the first of the four planets which make up the second group that we have thought it proper to constitute, in order to facilitate and simplify our explanation.

These four planets, in fact, possess a harmony of common characteristics, which permits us to class them together. Jupiter and Saturn are of enormous dimensions, greatly surpassing in size the other planets. They are attended by moons, or satellites; whereas the Earth is the only one of the first group of planets which is thus distinguished. Their years and seasons are of long duration, and their illumination by the Sun extremely feeble. Their days are very short, compared with ours. In fine, these planets turn on vertical axes, more fortunate in this respect than the planets that we have been viewing.

Jupiter is the largest planetary sphere of our solar system, fourteen hundred times larger than the Earth. It is 200,000,000 miles from the Sun; and, owing to this great distance, its year — that is, the time occupied in its revolution around the Sun — is equal in length to twelve of our years. Notwithstanding its colossal size, Jupiter turns so rapidly on its axis that in ten hours it makes a complete revolution; hence its days and nights are each only five hours long.

Much has been said about the brevity of Jupiter's days. In his "Picture of the Heavens," the German astronomer, Littrow (these Germans think of nothing but gormandizing), asks how the people of Jupiter order their meals in the short interval of five hours, and pities the ladies for the enforced brevity of their balls and parties. He rejoices, on the other hand, in the thought that the astronomers of Jupiter can see the most beautiful stars at mid-day, and with the naked eye, by reason of the faint illumination of that planet, which, owing to its distance from the Sun, receives twenty-two times less solar light than we do. A fantastic and illogical likening of Earthly conditions and customs to those of other planets — of which yet nothing can give us any idea, it is plain — always results, if we depend in making it exclusively upon what is in our own world.

It is true that Jupiter receives twenty-two times less solar light and heat than we do; but, as we shall see farther on, this is no proof that it is colder in Jupiter than in the Earth.

In compensation for its short nights, Jupiter has about himself four moons or satellites, which supply a steady light. This reflected light and very long twilights must render his nights hardly less brilliant than our days.

With the disadvantage of very short days, Jupiter has the inestimable advantage of perfect equality in the length of them and of his nights, and of the four seasons, on all his parallels. Of the obliquity of axis which produces in Mercury, Venus, Earth, and Mars the grievous inequality that we have noted in the duration of days, nights, and seasons in the same latitude, there is almost none in Jupiter. This planet turns on its geometric axis with hardly any inclination. Hence it has, like the planet Saturn, a kind of perennial spring; that is, a distribution of solar light and heat which operates in equal proportion along the same degrees of latitude.

Jupiter, then, has no vicissitudes of seasons, no abrupt and grievous alternations of cold and heat in the same place, like the Earth and Mars, Mercury and Venus. Its climate is uniform in every latitude, and the changes of the seasons hardly perceptible.

A gaseous atmosphere surrounds Jupiter, and clouds

float constantly above its surface, so that it is impossible to determine by the use of our telescopes its geographical configuration. We can only discern that the clouds are driven to the equator by the winds or swift whirlwinds: these tracts of clouds reach even to the tropics. There are, then, in Jupiter periodical winds — trade-winds. At the poles, which are more flattened than those of the Earth, the telescope enables us to see — as we saw in Mars, and as we know the fact to be on Earth — shining masses of ice. Cassini and several other astronomers declare that they have seen, in the equatorial regions of Jupiter, snow falling from the clouds and melting instantly.

Fig. 8. — Comparative magnitudes of Jupiter and the Earth.

Figure 8 shows the comparative magnitudes of the Earth and Jupiter, and also exhibits the nebulous appearance that Jupiter wears when viewed through the telescope.

According to the geodesic and climatological conditions that we have mentioned, the globe of Jupiter is admirably adapted for the home of living beings. Life to animals and men must be pleasanter in it than on Earth. Perennial spring reigns: excessive heat and extreme cold are unknown. The changes of temperature are nothing, because the axis of the planet is vertical. Spring being perpetual, flowers — or the ornaments, whatever they may be, of the vegetable kingdom — are always to be seen, new ones springing up constantly to replace those which have matured. The seasons are so long that the return of each is imperceptible, — a year of Jupiter being equal in length to twelve of our years.

We have said that Jupiter is 1400 times larger than the Earth; that is to say, the surfaces of each being measured, Jupiter's is 126 times greater than the Earth's. It results from this that the inhabitants of Jupiter can distribute themselves over a space 126 times more expanded than that of our globe, and can seek also a residence more suited to their tastes, or can please their minds and their eyes by far and easy journeys.

Placed 364,000,000 leagues from the Sun, Saturn is 734 times larger than the Earth. It occupies thirty years in its journey around the Sun. Its year, therefore, is thirty times longer than ours.

Like Jupiter's, Saturn's days are very short. In ten hours it turns on its axis: its days and nights are thus only five hours long. But eight moons or satellites, which accompany it, illuminate its nights, and compensate it, as Jupiter is compensated, for the brevity of its days.

The seasons of Saturn are each seven days and four months in length. In its long winters there may be seen, as in Mars and Jupiter, heaps of snow accumulating and remaining at its poles.

An atmosphere, distinctly recognizable through its effects, surrounds Saturn. It is easy to establish the fact that this atmosphere is periodically disturbed by equatorial winds; and alternate bright and dark belts, produced by the passage or absence of the clouds, are seen to furrow the middle of the globe.

In Saturn, therefore, as well as in Jupiter, there are trade-winds; that is, winds that blow periodically, like those of the equatorial seas.

Owing to its distance from the Sun, Saturn receives 260 times less light and heat than our globe.

The obliquity of its axis being almost nothing, its days are always equal to its nights. It has a perpetual equinox, its climate is uniform, and the variations of its seasons inconsiderable. As in Jupiter, spring endures for ever. This planet must therefore be as delightful for residence as is Jupiter.

Nature seems to have lavished its marvels on this fortunate globe. In its warm, equatorial regions must be found all those products that result from the activity of life. Rains, excited by the aerial currents, must stimulate vegetation, and produce all Nature's magnificences of growth. Ice gathered at the poles pours down an abundant tribute of flowing streams, while the clouds supply the basin of the seas. Thus there is a constant irrigation of this globe, a condition most favorable to the development of vegetation.

As if to enhance the brilliant spectacle of these natural adornments, Saturn enjoys a geodesic peculiarity which is shared by no other member of our solar system. An unaccountable exception among all known stars, Saturn is set in the centre of a ring of the same nature as its own, which surrounds it on all sides, at a distance of 800 leagues. This ring is enclosed by another, very much like it; and this in turn by a third. These are Saturn's Rings, so called by astronomers. This circular envelope is extremely thin (only 10 leagues thick), and is yet very broad (12,000 leagues wide). It is not fixed, but revolves about the globe which it surrounds.

Figure 9 shows the remarkable position of Saturn and its rings. It also indicates the comparative magnitudes of Saturn and the Earth.

Over the heads of the people of Saturn stretches always a luminous belt, a sort of eternal rainbow, which in different seasons is larger or smaller, and more or less distinctly visible, and on which the rays of solar light must produce indescribably brilliant effects.

The singular arrangement of the rings of Saturn illustrates the inexhaustible wealth of Nature, and the infinite variety of forms that the Creator has projected in the vast domain of the Universe. It should admonish us against our constant tendency to model on the pattern of the Earth other worlds than those we know.

Fig. 9. — Saturn and the Earth, their magnitudes compared.

We said just now that Saturn is attended by moons, or satellites, intended to light up his nights. Figure 10 shows the respective distances of these moons and their positions.

We shall say but little of Uranus, a planet which is only 82 times larger than the Earth, and which yet revolves at a distance of 732,000,000 leagues from the Sun, and fills 84 years in his journey around that centre. Uranus, by reason of its distance from the Sun, is favored with 360 times less heat than the Earth receives. It, too, has eight satellites.

Fig. 10.—Saturn and its satellites.

In Figure 11, the magnitudes of Uranus and the Earth are contrasted.

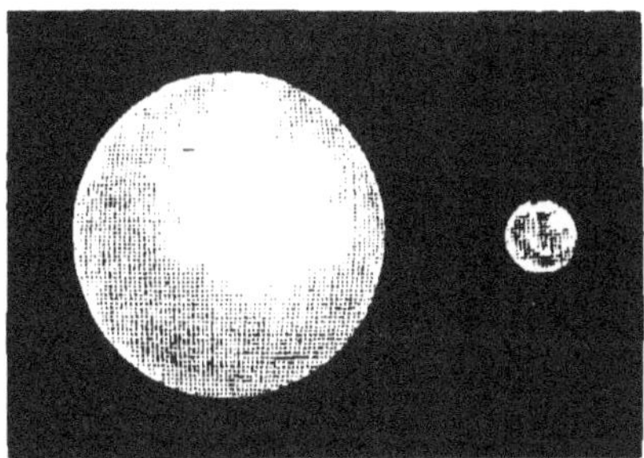

Fig. 11.—Uranus and the Earth, their magnitudes compared.

Its immense distance, together with its diminutive size, makes it almost inaccessible to our astronomical instruments, so that we can make no positive affirmations as to its geographical configuration.

For the same reason we shall say even less of the last star of our system, Neptune, which was discovered in this century by M. Leverrier, by the aid of the Calculus,—a discovery which has supplied the most striking proof ever obtained of the great value of mathematical science. Neptune (Figure 12) is so small and so distant that probably unaided scrutiny of the heavens

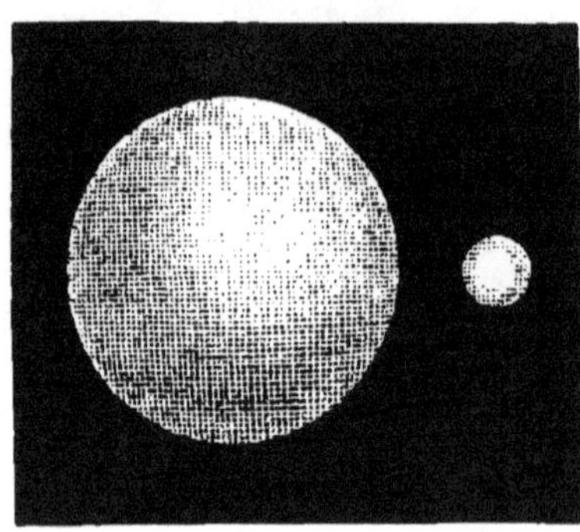

Fig. 12.—Neptune and the Earth, their magnitudes compared.

would never have discovered it. In this case mathematical analysis was more powerful than the telescope.

It would be impossible for us to give such particulars of information as we have sketched about the preceding planets, with reference to a star only 105 times larger than the Earth, *which moves at a distance of* 1,150,-000,000 *leagues from the Sun*, and whose year is 164 times as long as ours; so that, estimating time according to the Neptunian chronology, we should reckon, instead of nineteen centuries, only twelve years of the Christian era. All that we can say of Neptune, besides this, is that it terminates the domain of our visible world.

Yet we cannot affirm that our solar world ends at these bounds. It is true the reach of our glasses stops at Neptune; but that planet is not, be assured, the frontier of the Sun's empire. We know, indeed, that the comets come back to us from a journey through space of 32,000,000,000 leagues, as is shown by their geometric curve. Thus the distance of 1,150,000,000 leagues, which separates Neptune from the Sun, by no means represents the confines of our solar world: it expresses only the limits of the compass of our telescopes.

The exhaustive examination we have made of the question whether or not the planets nearest the Sun — Mercury, Venus, the Earth, and Mars — are habitable enables us to abridge materially our discussion of the same subject in connection with the second group of planets of which we have just given some astronomical and climatological *data*. What we said in Chapter V. as to the probable existence of organized life, plants and animals, in the planets of the first group, applies equally to those of the second. We said, in that chapter, that wherever there is air, water, and a moderate temperature, organic life must be. These conditions are found in Jupiter, Saturn, and Uranus. There is air in those planets; for their atmospheres betray themselves by optical phenomena of the deflection of light, as well as by the aurora and twilight. There is also water; for clouds are plainly visible at the equators of Saturn and Jupiter, and ice shows itself at their poles in the shape of white spots, which melt as winter wanes in these parts. Consequently liquid water must exist on the surfaces and in the declivities of these planets; it must fill the vast basins that form seas, and flow through valleys to the interior of the continents as rivers and streams. Besides, there are regular and irregular winds in Saturn and Jupiter, which sweep and mingle the strata of air, and carry them rapidly from one region to another. On Jupiter as on Saturn, we see clouds driving along the equator, and ridging that vicinity with bands far brighter than the tropical and ultratropical parts. The rapid hurrying of the clouds proves to the scientific observer that periodical winds (trade-winds) blow at the equators of these two planets, as on the Earth. Under the influence of these winds, evaporation must be very rapid at the surface of the soil; vapors must rise and form new clouds. These clouds resolved in turn into rain, the globe also is washed by fertilizing showers. A continual exchange is also established between the water which ascends in vapor and the clouds which descend in rain, and effects that watering of the earth which is as essential to vegetation as is exercise to animal life.

Animal life, supposing it to exist in these planets that we are surveying, must be easier and more delightful than in our globe. As the planets revolve on vertical axes, the climate is uniform in every latitude, and temperatures diminish only according to the curve of the globe; that is, very gently. The changes of season are almost imperceptible in Jupiter and Saturn, and undoubtedly in Uranus also. These planets know no sudden changes of weather, no unexpected transitions from cold to heat, from wintry air to that of spring or summer, such as happen on Earth, in Mercury and Venus. They enjoy a kind of perpetual spring. Such a privilege is plainly very favorable to the regular functions of animal and vegetable life.

The only objection that can be urged against the habitability of Jupiter, Saturn, and Uranus, is founded on their temperature: we can hardly imagine life in lands so feebly lighted, so meagrely heated by the Sun.

It is certain that that radiant luminary, which, as the poet Lefranc de Pompignan says, —

> Pours torrents of light
> On his obscure blasphemers,

is not very dazzling on the Jovian or Saturnian planes. Jupiter gets 22 times less sunlight than the Earth, and Saturn 260 times less. Twenty-two times is a trifle, when we consider that in solar eclipses, when five-sixths of the Sun's disk is obscured, our light and heat are hardly perceptibly diminished. But 260 times less is more significant, and in this case we must apply the considerations on which we are going to enter.

In arguing about the habitability of the planets, we must divest ourselves carefully of every recollection, every prejudice, as to what is going on about us. We must not limit the resources of Nature and the infinite variety of her appliances to the few specimens that are within the range of our vision. Animals and plants in the extraterrestrial globes are certainly not made on the same plan of organization, have not the same forms, as their correspondents on the Earth. Their structure is adapted to the medium in which they live. The men of Venus and Mercury are made by Nature to resist heat, as those of Jupiter and Saturn are made to endure cold, and those of the Earth and Mars to live in a mean temperature; otherwise they could not exist.

We must not, however, be frightened by figures. No doubt Saturn receives 260 times less heat than the Earth; but we must remember that this globe is surrounded by an atmosphere. The atmosphere is the savior, the palladium of the planets. By its mere presence it regulates their temperature, — maintains or mitigates it, according to their needs. It is a beneficent Proteus which looks out for every thing. According to its composition, its density, its capacity for caloric, or its conductivity, the atmosphere can preserve the heat of the Sun accumulated on a planet, or, on the other hand, disperse it swiftly through its mass, to prevent an accumulation at one point.

Let us try to show how infinitely useful is the atmosphere that surrounds our Earth.

The atmosphere serves our globe in the capacity of a hot-house, that retains and preserves the heat and light from the Sun. Without its aid, these would be lost. Reflected directly from the rocks and bare spots, the light and heat would be lost in celestial space. This happens on the peaks of the American Andes, on the summits of the Alps and the Himalayas. On these mountain-tops the heat of the Sun, not finding air enough to retain it, distributes itself through space, leaving these heights a prey to the rigors of excessive cold, which is fatal to all organized life.

Without the atmosphere, there could be no liquid on the earth: water could not be kept there. In fact, to prevent a liquid from passing into vapor, some gas, or some other liquid, must weigh upon it. This is the office of the air on our globe. By its pressure it prevents water and other liquids from vaporizing. Suppose that the earth were suddenly stripped of its gaseous envelope: instantly all the waters of the seas and rivers, all that the Earth drinks in, all that impregnates living bodies, would exhale in vapor, losing itself in space, and the Earth would become the house of death.

This is plain enough to one who has seen the beautiful experiment of placing a drop of water under the glass pump of a pneumatic machine. After several strokes of the piston to raise the water that fills the pump, the water begins to boil, though cold, and, dissolving into vapor, disappears. This is the image in miniature of what would happen throughout the whole globe if, by some chance, the atmosphere that presses on its surface should be driven away by the magical piston-stroke of some gigantic experimenter.

The atmosphere also produces sounds and reports. Without it the Earth would be the most doleful of places, the abode of absolute silence. Noise and stir are the essence of life. A land where no noise resounded, where no sound struck the ear, would be a most dreadful tomb: men and animals would be deaf and dumb.

Sound is produced only by the agitation of the air caused by the vibrations of bodies. The air lacking, there is no sound. Every one has witnessed this beautiful experiment: Place under the receiver of a pneumatic machine the striking parts of a clock, which sound clearly under the receiver; take away the air from the receiver by a few piston-strokes, and the bell no longer sounds; the hammer strikes, but no sound is heard. It is because air, the soul of sound, is gone. But let in a little air by opening the valve, and a faint sound will begin to be heard. Fill the receiver with air by opening wide the valve, and the bell will resound distinctly. This experiment goes farther than a long process of reasoning to make us understand that air is the vehicle of sound. By consequence, an atmosphereless planet would be abandoned to absolute silence.

While it is an indispensable condition to the exercise of our sense of hearing, the atmosphere is also necessary to vision. It is the air that distributes the light over a wide extent of space. Without the intervention of the air, the Sun's rays falling on any surface whatever would be directly reflected, lighting up only the points touched, or those on which it may be reflected. It is the air that scatters the luminous rays in all directions, and makes half-illuminated shade possible. Without the air, the only objects visible would be those reached by the Sun: elsewhere darkness would reign. There would be no twilight, or partial obscurity:-we should be sensible only of the night or the dazzling rays of the Sun. It is well known that twilight is occasioned by the gradual distribution of light: though the sun has sunk below the horizon, light remains; this is because the air has retained and slowly scattered in every direction the last rays of the setting source of light. The same principle is witnessed in the dawn. The Sun would appear on the horizon abruptly and unannounced, we should pass at once from the darkness of night to the sudden glare of day, if the air did not retain and distribute over a wide expanse the rays of the rising orb.

Without an atmosphere, we should have in place of the blue of the heavens, but a black and crude surface. This heavenly vault, so purely and serenely blue, is nothing but air. Blue is, in fact, the color of the air; and its color cannot be distinguished, except when seen through a great mass of this fluid, as is the case when we look into the heavens, — that is, into the mighty stratum of air which presses on the Earth. Without air, consequently, the heavens would be colorless: we should see in them only a black and monotonous expanse.

Let us add that the atmosphere — which, in its physical qualities alone, is, as we have seen, essential to the exercise of the functions of life — has an important bearing, through its chemical character, on the vital functions of plants and animals. From the air vegetables draw a great part of their nourishment, or, to phrase it better, of the elements of their sustenance. From it also men and animals obtain the oxygen needed for their breath.

We simply express an incontestable truth when we say that air is life in these planets, and without it there would be on their faces only silence and death.

In Chapter VI. we cursorily surveyed the moon,

although she seems properly to be closely connected with the earth. We have restricted our mention of this satellite to a few lines. And why? Because the moon is an atmosphereless star. Whatever cause may have stricken her with this terrible disaster, whereby she neither has nor ever had an atmosphere, it is certain that she now lacks this gaseous envelope. Hence the moon is uninhabitable: she can give a home to no living creature; and upon this ground we have thought it proper to exclude her from special mention in this work.

But it is otherwise with the planets. Jupiter, Saturn, and Uranus have atmospheres, as well as the planets nearer the Sun. Their atmospheres mitigate the inconveniences which result from their great distance from the Sun, acting like muffs to retain and husband the heat that they receive from the central source. However feeble the conductivity of heat of this atmosphere may be, its capacity for containing considerable heat and its somewhat greater density must suffice to enable Jupiter, Saturn, and Uranus to retain solar light and heat, and prevent them from losing their caloric by radiation, except very gradually. In the short nights of these planets, cold has not time to become very severe; hence, in fact, the temperature of these planets so distant from the Sun is not, perhaps, much lower than that of the Earth or Mars.

An hypothesis that we are going to venture upon will perhaps enable us to reconcile the relative distance of the planets from the Sun with the existence of an almost uniform temperature in them all, and make it appear that Mars, which adjoins the Sun, is not much hotter than Uranus, which is 732,000,000 leagues from it.

Seen from either planet, the Sun presents decreasing proportions; and Figure 13 shows exactly the comparative size which it presents to the inhabitants of each planet. Its disk diminishes as the distance of the observer increases. The heat transmitted by the Sun diminishes in proportion to its apparent size.

All the planets, we have said, have an atmosphere. Suppose that the power of the atmosphere of each planet to absorb light and heat should increase in order, from Mercury to Venus, from Venus to the Earth, from the Earth to Mars, from Mars to the Asteroids, and even to Neptune, at the extremity of the solar system. Suppose, in other words, that the nearer a planet is to the Sun, the greater is the ability of its atmosphere to absorb caloric. On this supposition, Mercury and Venus will absorb almost the entire solar heat, while their temperature will be moderate. The Earth and the Asteroids receiving less light and heat than Venus and Mars, but their atmospheres having greater absorptive power, their temperature will not differ greatly from that of Mars and Venus. The same reasoning is applicable to the whole group, even to Uranus and Neptune, in which, despite their distance, their atmospheres having no absorptive power, all the solar heat and light that reaches them is utilized; so that Neptune and Uranus cannot be much colder than Mars or the Earth.

Thus in our moderator-lamps the light is rendered

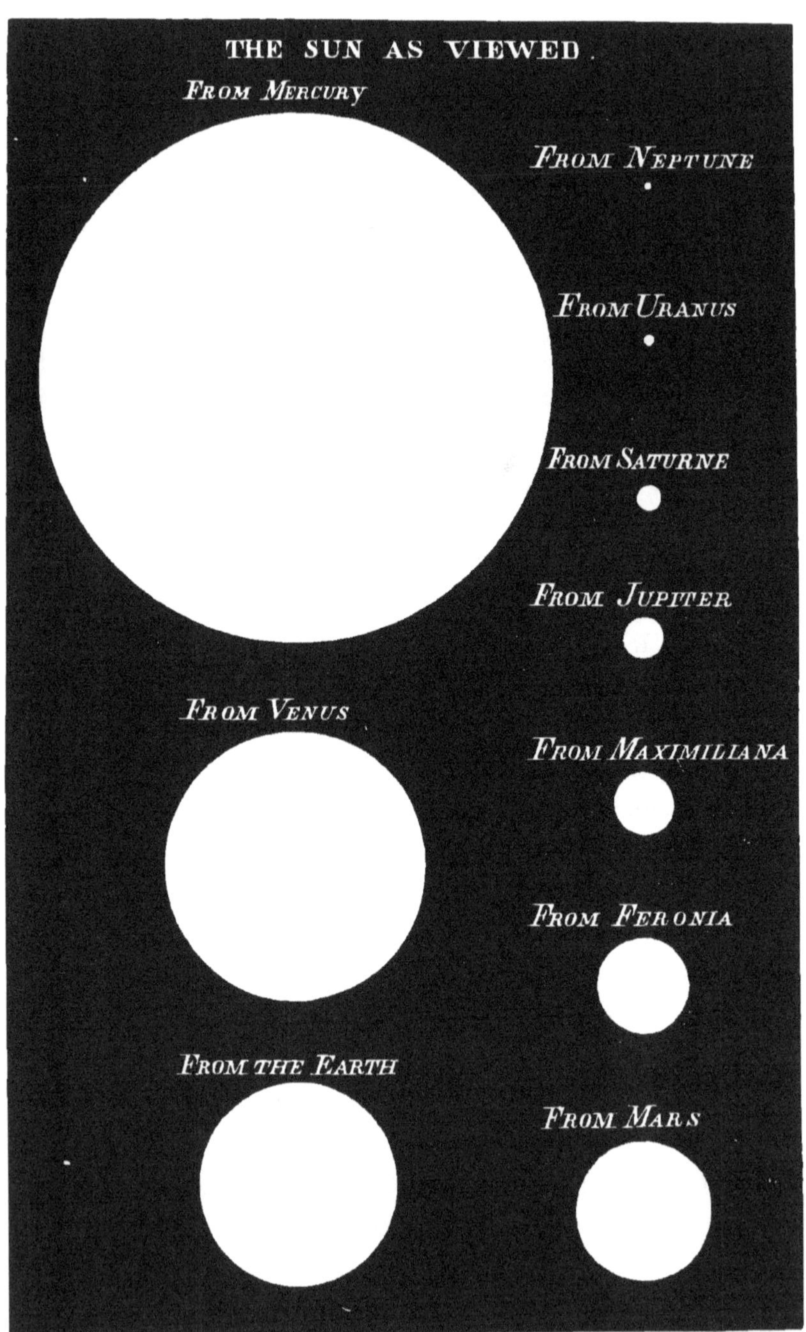

Fig. 13. — The Sun as seen from the several planets.

of equal intensity by an ingenious artifice of this effect: that in proportion as the moving spring loses its force and presses less vigorously on the oil to raise it, the tube by which the oil reaches the burner expands, and permits more of the illuminating liquid to pass to the wick.

If, then, our hypothesis is admitted, that the absorptive power of the atmosphere of the planets increases in proportion to their distance from the Sun, it results that the temperature of them all must be nearly uniform.

There is another influence which tends to maintain the heat of the planets, notwithstanding their distance from the Sun. It is their own temperature; that is, the temperature which they have independently of the Sun. It is now known that the planets were originally in a liquid state produced by heat. Only by cooling have they become solid and able to offer a home to organized life.

This individual heat is still preserved in their centres, where, if not on their surfaces, it is quite perceptible. At several hundred metres below the surface, the Earth is exceedingly hot. There is no reason to doubt that Jupiter and Saturn, owing to their enormous size, were longer in cooling than the Earth, and that their surfaces may be still impregnated with sufficient caloric from this intense source of heat to supply the deficiency of the solar provision. The central fire, as it is called, can maintain on the surface of these planets the temperature essential to life.

We endeavor to meet carefully the objection drawn from the low temperature of Jupiter, Saturn, and Mars, as we have met the argument founded on the high temperature of Mercury and Venus. We do so because this is the argument that is oftenest offered in denial of the hypothesis that living creatures exist on the distant planets. The reader must now be convinced of its weakness. All the arguments based on the excessive distance of the Sun, as well as those based on its too close proximity, and which seem to deny the possibility of life on some planets because it would be roasted, and on others because it would be frozen, are put to naught by a correct philosophy. It is a great mistake to make the mean temperature of the Earth serve for the thermometrical zero of the worlds, and to employ the Earth as an absolute term of comparison with the other planetary globes.

We must conclude, therefore, that in the farthest, as in the nearest planets, life may be as regular and as simple as on our globe, and that Jupiter and Saturn are as habitable as the Earth.

To sum up, whether Nature creates peculiar organisms adapted to the physical and climatological conditions of each region; or whether she mitigates by the composition of the atmosphere the evils which would result from excessive remoteness from the Sun, or from too close proximity to it, — we may be sure that life exists in those planets; that all the globes that revolve around the Sun together with the Earth are covered with vegetation, and wholly peopled with their own *fauna.*

To maintain the contrary would be to hold that the Earth is an exception to the rule of the Universe; to declare that, while it is the always open theatre of vital activity, the worlds, of which it is one, are only vast and useless deserts.

Before science, with the most rigorous evidence, had demonstrated the existence of globes at all like ours; before the telescope had enabled us to see — we might almost say, to touch — these far worlds, we could believe in the exclusive pre-eminence of the terrestrial globe; we could say that the moon and stars were created solely for its decoration. But to-day it would be impossible to maintain such a theory. What, the Earth, that represents only a grain of dust lost in infinite space, shall it be the only seat of life; and shall planets a hundred times, a thousand times, fourteen hundred times larger, be only a vast grave, the one nothing in the Universe, the one empty edifice in the economy of Nature? Is life on our globe, — that insignificant atom, — to be heaped up, pressed down, and running over, filling every space, so that not a corner of its surface is empty, while in the rest of the Universe not a sign of life is discoverable? What, because our eyes are too feeble, our instruments too imperfect to recognize the men and animals of the planets, we must suppress the living creatures of these kingdoms of space; we must rob of their natural adornments those magnificent spheres which move, like ours and in the same time with ours, around the orb of day; we must see in those majestic spheres that roll over our heads nothing but gloomy solitudes; we must condemn all the planets to the condition of inert blocks, doomed to eternal barrenness; we must make of them humble vassals, who bow and retire before the paltry majesty of our wretched globe; we must conclude that God exhausted his omnipotence in the embellishment and decorations of this little Earth, and that he has set aside, as unworthy of his notice, the whole remainder of the Universe! The soul revolts at and resents the mere statement of such a doctrine!

Science has shown, as we have just seen, that the natural, physical, and climatological conditions of the Earth and the other planets are identical. In those planets, as on the Earth, the Sun shines and disappears, yielding place to night; and cold and darkness succeed to heat and light. In them, as on the Earth, the rich carpet of herbage covers the plains, and luxuriant woods cover the mountains. Rivers flow majestically on, bearing to the bosom of the seas their unfailing tributes of the rains or of the fountain of polar ice. Winds blow regularly or irregularly, and purify the atmosphere by mingling their strata, charged in different degrees with the product of the evaporation of their soil. In quiet nights dwellers in these planets see the same heavenly spectacle that delights our eyes,—the same constellations, the same celestial vistas. They have panoramic views of the planetary globes with their followings of faithful satellites, and of luminous stars shining like gently-brandished torches. Once in a while there is a sudden luminous trail which furrows the heavens like a flash of silver: it is a star that shoots, and drops into the depths of space. Again, it is a comet with a beautiful tail, that comes to bring us news from worlds millions of miles away.

And all this must count for nothing! The Sun looks upon these worlds to illuminate nothing, to vivify nothing, to give nurture and life to no atom of organic matter! These beneficent rains can wash nothing; there cannot be a tuft of grass to drink in the blessing dew. The seas that cover a great part of the surface of these planets, as they fill the bosom of our globe, must be empty, uninhabited; not a living thing can they hide in their waters! The winds must blow from the equator to the poles, — here periodically, there irregularly, — and to no purpose! The brooks must glide down the mountains and flow into the folds of the valleys; streams and rivers traverse the plains, rivulets furrow the woods, — and all these must wash nothing, fertilize nothing, refresh nothing! Breezes and rains, winds and waters, — all these for deserts! Not an animal, not a plant must enjoy these blessings of Nature! The radiant face of the starry heavens; the glowing light of day; the marvellous perspective of verdant landscape; the grand horizons of the mountains, and the limitless expanse of prairies, — there must not be an eye to gaze upon these; not a soul to enjoy their beauty, not a mind to admire them, to comprehend them, and to thank the Creator for his goodness! Ah, this is blasphemy! The idea of a world that accomplishes its solitary task, without a living being that may be called to share it, — of a machine complicated and skilful, that turns in a void, endless, objectless, useless, — this idea is monstrous. It is almost an insult to the all-powerful Deity; for God cannot have created any thing superfluous.

I may be deceived, but it seems to me that the demonstration of the existence of planetary worlds like ours ought to suffice to prove conclusively that there are in them beings which reason, souls which feel, and intellects which think. It seems to me that the beautiful, the grand, the sublime would not be were there not intelligences to comprehend and souls to feel them.

To sum up in one word all the arguments and all the facts that we have gathered, let us say that, if the planets are habitable, they are inhabited.

CHAPTER VIII.

There are in all the Planets Vegetables and Animals, as well as a Higher Type, the Planetary Man, corresponding to the Terrestrial Man. The Process of Creation of Organized Life on the Earth must be the Same in the Planets. The Successive Order of Appearance of Living Creatures on our Globe. The same Succession must exist in Every Planet. The Planetary Man. The Planetary, like the Terrestrial Man, is transformed after Death into a Superhuman Being, and passes into Ether.

WE have shown that organized life exists in all the planets. But is the life that exists in the far worlds like that of plants and animals on the Earth, and is it accompanied by a superior type, like the man of Earth? Let us examine this point.

Guided by analogy, the sole means of investigation that can be used where observation is impossible, we admit that what has happened on the Earth since the time of its formation must have happened on all the other planets, its congeners.

It is well understood, at this day, how the vegetable and animal creations have appeared and succeeded each other on our globe, since its creation. The Earth originally was only a body of gas and burning vapor which flowed around the Sun. This mass, cooled in its journey through space, took at first a liquid state, passed then to a pasty consistency, and at last became solid under the prolonged influence of cold. The consolidation began at the surface, because the circumference of a sphere is most exposed to frigerating influences. Then the water and vapor that floated about the consolidated globe were condensed, and, falling in scalding rains upon the stiff soil, formed the first seas.

That the Earth was originally in a liquid or semi-pasty state is shown by taking a plastic sphere, — a lump of soft clay, for instance, — and making it revolve rapidly on its axis. A swelling of its middle parts and a flattening at the poles of its axis will be observed: this is the effect of centrifugal force engendered by the rotatory motion. Now the Earth is flattened in this way at its poles, and a little swelled at the equator.

What happened on the Earth must have been paralleled on all the other planets at the time of their formation. They were composed, at first, of gas and vapors, which in cooling became liquid, then pasty, and at last solid. The congelation especially affecting their surface, they began to present a solid bark, or external envelope: this was the soil of the planet. On this hard soil the liquid resulting from the condensation of the vapor of water fell, and remained; and thus were formed the first seas of the planets.

To any who question this theory we could exhibit the globes of Saturn and Jupiter, whose poles (as has been stated on p. 62) are even more flattened than those of the Earth. What is the cause of this greater oblateness? It is that these planets turn on their axes much more rapidly than does the Earth: our days are 24 hours, while those of Jupiter and Saturn are only 10 hours long. A greater rapidity of rotation must produce a greater depression at the poles. This geometric result clearly demonstrates the propriety of our comparison of the Earth and the other planets, in investigating their origin.

In the yet warm waters of the basin of the seas appeared the first living things that existed on our globe. Animal life began in the water. Zoöphytes and mollusks first became visible: this we know, because zoöphytes and mollusks, together with some articulates, compose the animal remains found in the transition rock which comes next to the primitive.

As to the first-grown vegetables, those whose prints are found in the same transition rock are acotyledonic, cryptogamic; that is, mosses, algæ, and ferns.

When the Earth had grown a little cooler, phanerogamic plants (monocotyledons and dicotyledons) appeared on the continents. Many kinds of vegetables were created simultaneously; for the flora of the secondary soil is exceedingly rich and various.

The process was similar with animals. To the zoöphytes, mollusks, and fishes that are found in the transition rock, succeeded, in the secondary, reptiles, which appear in the sea as well as on the land. Then came those monstrous saurian reptiles, whose extraordinary forms as well as their colossal size strike us to-day with astonishment, almost with terror. Then the gigantic mosasaurus ravaged the seas, the terrible ichthyosaurus spread terror among the inhabitants of the waters, and the huge iguanodon depopulated the forests. The secondary soils, filled with their bones, prove to us that in the secondary epoch reptiles held the highest place in creation.

Later, the atmosphere having been purified, birds began to cleave the air. In tertiary soils are found remains of several kinds of birds, which, not being found in anterior soils, sufficiently show that during the tertiary epoch birds first appeared on the Earth.

Later still — that is, after the tertiary epoch — mammiferous creatures first entered life.

It should be remarked that all these animal species do not mutually replace each other, that one kind does not always exclude another. Many ancient animal species existed after new ones had appeared. We could name such groups: as the lingule (mollusks), the coral (zoöphyte), the oyster (mollusks), among animals; and, among vegetables, algæ, ferns, lycopodia, which came upon our globe after the first periods of the organized animal kingdom, and which have not ceased to exist down to our day.

Not until the last epoch in the history of the Earth — that is, in the quaternary — did man appear, the highest type of the living creation, the last term of organic, intellectual, and moral progress; the crown, on our Earth, of the visible edifice of Nature.

To-day man lives in company with animals which saw the light in the quaternary epoch, and with very many other species of mammiferous creatures which were created in the tertiary epoch.

Such, in brief, has been the progress of the vegetable and animal creation on the Earth.

These diverse phases which have followed the development of the vegetable and animal kingdoms, these improved organic species which add themselves to each other and adjoin the superior type that is called man, must, we believe, be paralleled in the other planets of our solar world. The physical constitution and climate of these planets are almost identical with those of our Earth; and there is no reason for believing that this progress of growth should be different on Mercury, Venus, or Jupiter, from what it was on our Earth.

There must have been, we think, on these planets a successive appearance of vegetables and animals, whose types were improved from age to age. The plants and animals of Mercury, Jupiter, and Saturn, and the others, were not, surely, identical with those of the Earth. Perhaps they resembled them in no respect; but all, in their successive appearance, obeyed the principle of progress and improvement. Beginning in the scalding waters of the first seas, life showed itself on the continents. Animals of aerial organization lived on these continents: their species gradually improved on their type; and at last, as the final term, appeared in these planets a complete being, superior in organization, in intelligence and in sensibility, to all the rest of the animal creation which populated the globe.

This superior being, this last round in the ladder of the living creation of the planetary worlds, and which corresponds to man on Earth, let us designate as planetary man.

In all the planets, then, as on the Earth, there are men, together with animals of a lower type.

According to the opinions expressed in the beginning of this work, terrestrial man undergoes, after his death, a glorious metamorphosis. Leaving here its wretched material garment, his soul darts into space, and goes to be incarnated in a new being, whose type is infinitely superior, in moral perfections, to that of our poor humanity. He becomes what we have called a superhuman being. If this proposition holds good as to terrestrial, it must be equally applicable to superhuman man. Hence the superhuman being must come not only from the Earth, but from the other planets.

They come, then, from human souls which have lived perhaps on Earth, perhaps in Mercury, in Jupiter, in

Venus, in Saturn, &c. And as the superhuman being who comes from Earth enters the ether that surrounds it, so the planetary man going from Mars, or Mercury, or Jupiter, passes into the ether which surrounds his own planet, is there incarnated in a superhuman body, and lives in those regions of ether which adjoin the planet whence he came.

Thus the principles that we have laid down for terrestrial humanity are generalized, and become applicable to all planetary humanity. Not from the Earth alone spring the souls that go to be incarnated in new beings, in the bosom of ethereal space: they come from all the globes that, with the Earth, make up the retinue of the Sun.

All these superior beings float on beds of ether, which in each planet lie next to its atmosphere.

We are now going, rash though it may be to touch on such a subject, to try to convey an idea of the bright creatures that swim in those mysterious and sublime regions, in that empyrean which is hidden from our eyes. In other words, we are going to try to comprehend the attributes, the form and qualities of the superhuman being, whether he is of terrestrial or planetary origin.

CHAPTER IX.

What are the Attributes of the Planetary Man? His Physical Form, his Senses, Degree of Intelligence, and his Faculties.

LIKE the human, the superhuman possesses three elements of "the aggregate;" namely, the body, the soul, and the life. In order to form an idea of him, we must examine separately each of these three.

The Body of the Superhuman Being. — It is possible, perhaps, to conceive of a bodiless superhuman being; to imagine that the soul, purely spiritual, constitutes the happy being who soars in ethereal space. It is not thus, however, that we conceive of him. This absolute immateriality must, we think, belong only to a being higher in the moral hierarchy than the mere superhuman, and of whom we shall speak farther on. We believe that the inhabitant of ether has a body; that the soul, parting from its earthly residence, goes to take up lodging in a body, as it did here below. But the body must possess qualities infinitely superior to those which belong to the human tenement.

Let us inquire, first, what can be the form of this body? The painters of the Renaissance, whom modern artists have imitated, give to angels the form of a man, young and beautiful, with white wings with which he flies through the air. This image is both poetic and gross: it is poetic, in that it responds to the idea that we must form of the shining beings who hover in the upper ether; it is gross, because it invests a creature far above man with the physical attributes of a mere man, which is unwarrantable.

These painters who, following the example of Raphael, depict cherubim as having the head of a child, furnished with two wings, express the same thought, but in a manner more profound. In suppressing the bulk of the body, and restricting the seraphic being to a head, the seat of intelligence, they seem to imply that in the Christian angel the spiritual nature infinitely predominates over the natural.

We cannot be expected to delineate the forms of superhuman beings. We can only say that, ether being a fluid exceedingly fine and rarefied, the superhuman being, in order to float in it, must himself be indescribably light, and must be composed of substances wonderfully subtile. A thin material tissue animated by life, a transparent and vapory cloak of living matter, — that is our idea of a superhuman being.

How does the body of this being sustain itself? Must it, like men and animals, eat to live? We reply confidently that alimentation, that necessity which tyrannizes over men and animals, is not imposed upon the inhabitants of ether. Their bodies must be repaired and sustained by the simple respiration of the fluid in which they are immersed; that is, of ether.

It should be remarked that the necessity of eating holds an important place in the vital economy of animals. Many of them, especially those which live in the water, are compelled to eat incessantly, under penalty of death by inanition. In superior animals, the necessity of eating and drinking is not so imperative; for the breathing function brings to the body, by absorbing oxygen and a little azote, a modicum of those reparative elements which supply the place of alimentary substances. This point of superiority is quite perceptible in man. Our breathing is a function of great importance, and helps materially to keep our organs in repair. Oxygen, which our blood borrows from the air in respiration, enters largely into our nourishment. Among birds, whose breathing function is very active, and whose organs connected with it are large, the oxygen inhaled does a like service in nutrition, and is substituted for a great quantity of food.

We believe that in the superhuman being the respiration of the ethereal medium suffices to nourish the material body, and that therefore he is spared the necessity of eating and drinking.

I am not sure that the reader has formed a clear idea of the consequences which must follow the absence of the necessity of alimentation in the beings in question. These will be understood upon carefully considering that it is the obligation to look out for bodily nourishment that occasions the miserable condition of animal life. Obliged incessantly to seek for means of subsistence, animals are abandoned to this brutalizing preoccupation: from this result their passions, their quarrels, and their pains. It is the same with man, though in a less degree. The necessity of providing for his sustenance each day — of procuring his daily bread, as we commonly say — is the great cause of the fatigues and suffering of the human species. Suppose that man could live, grow, and sustain life without eating, the air he breathed sufficing to keep his organs in repair: what a revolution in human society would this ability represent! Hateful passions, wars, rivalries, jealousies, would vanish from the face of the earth. The age of gold, dreamed of by the poets, would be the certain consequence of this organic change.

This benefaction of Nature, denied to man, is certainly vouchsafed to the superhuman being. We must conclude, therefore, that the bad passions which are our sorrowful heritage must, in the same way, be unknown in the home of these favored beings. Freed from all care of seeking food, living and sustaining their strength by mere respiration, — an act involuntary and unwittingly performed (as the circulation of the blood and absorption go on in man), — dwellers in ethereal space must yield themselves wholly to quiet and gentle sentiments, and to no impressions but those of bliss and unalloyed serenity.

The forces of our body are speedily exhausted. We cannot exercise our organs beyond a certain time without feeling fatigue. To transport ourselves from place to place, to lift burdens, to walk, to ascend or descend, we must expend strength and soon pause from weariness. Thought can work only for a limited time: soon the attention fails, and thought is suspended. In fine, our corporeal machine, so admirably constructed, but so delicate, is subject to a thousand derangements that we call maladies.

Of the painful sensation of fatigue, and the continual menace of sickness through the disturbance of our organs, the inhabitants of ether know nothing. For them rest is not a necessity after exercise, as it is for us. The body of the superhuman being, insensible to fatigue, needs no repose. Unembarrassed by the thousand movements of a complicated machine, he lives and survives by the pure force of the life that is in him. Probably the inspiration of ether is his single physiological function; and that it is possible to exercise it without numerous organs is proved by the fact that in all of a whole class of animals, — the batrachians, — the mere bare skin constitutes the whole machinery of respiration.

The extreme simplicity that must appear in the body of the superhuman being will be understood if it is admitted that respiration is the only function which he has to exercise. The numerous and complicated apparatus and organs in the bodies of men and animals are intended to carry on the functions of nutrition and reproduction. These functions being dispensed with in the beings in question, the body must be by so much lighter. Its only duty must be by respiration to preserve and sustain the faculties of the soul: all its energies must combine for the exclusive end of spiritual nourishment. We reasonably admire the ingenious mechanism of the body of man and animals; but if human anatomy discloses wonders, marvels of foresight, intended to insure the preservation of the individual and his reproduction, how much greater marvels should we not find, if it were possible for us to seek them, in the organization of the superhuman body, in which every part is designed to sustain and perfect the soul! What astonishment should we feel in understanding the use and object of the different parts of this glorious structure, in tracing the relations of resemblance or of origin between the living human and superhuman economies, so as to discover those which may exist between the organs of a superhuman being and those which in another life must belong to a being still higher, yet the same, animated anew in glory and perfection!

The peculiar organization of the being we are describing must give him the power of transporting himself very rapidly from place to place, and of compassing great distances with extraordinary speed. Since thought in us poor humans leaps over space, and in the twinkling of an eye travels from one end of the globe to the other, we may believe that the body of superhuman beings, in which the spiritual principle dominates, must possess the marvellous power of passing from point to point of space with a celerity which we can measure only by that of electricity.

The superhuman being, who needs neither to eat nor drink nor rest, who is always active and alert, has also no need of sleep. Sleep is no more necessary to him for the repair of his forces than is nourishment to create them. We know that the obligation to sleep deprives man of one-third of his life. He who dies at thirty years of age has really lived but twenty: he has slept the other ten. What a paltry idea does not this give us of the condition of man! But whence arises this necessity of sleep? It is because our physical forces, exhausted by exercise, must restore themselves by quietude and cessation from activity, by a brief suspension of nearly all the functions of life, by a kind of temporary death. In sleep, man prepares and gathers the strength that he will have to expend when he awakes. He devotes the night to this recuperation of his forces, as much in conformity to the general rule of Nature as in obedience to the customs of civilization. But it is probable that, in the superhuman being, these forces are inexhaustible, and need not for self-reparation that state of slumber, which is one of the severest tyrannies of human life. Every thing leads us to believe that the being who comes next to us in the hierarchy of Nature is always awake; that the word "sleep" has no meaning for him.

It must be remembered, moreover, that night is unknown to these ethereal beings. With us night and day succeed each other, in consequence of the rotation of the Earth on its axis, which deprives the Earth of the light of the Sun during half of its revolution. This rotatory motion affects our atmosphere; but that does not extend to a very great distance. On the ether, above our atmosphere, this motion has no influence. That fluid mass remains motionless, while the Earth and its atmosphere revolve on their axis. The superhuman beings, who (according to our system, properly understood) inhabit the planetary ether, are unaffected by this rotation. They see the Earth turn, so to speak, under their feet; but, themselves placed far above the motion, they never lose sight of the central Sun. Night, we repeat, is an accidental phenomenon peculiar to the planets alone, whose hemispheres are alternately illuminated by and turned from the Sun; but the rest of the Universe knows no night. The superhuman beings who hover in the realms above the planet always see the luminous orb, and their blissful days roll on in a sea of light.

Passing now to the senses which must belong to these beings, we say, —

First, that they must enjoy the same senses that are ours, but that these must be infinitely more exquisite than ours. *Second*, that they must possess special senses that are unknown to us.

What are these new senses that the superhuman being enjoys? It would be impossible to answer this question satisfactorily. We know nothing of any senses except so far as we have ourselves employed them, and no man could divine the object of a sense which Nature had denied to him. If you try to give a blind man an idea of the color red, he will answer you, "Oh, yes! it is loud, like the sound of a trumpet." If you try to make a man born deaf comprehend the tunes of a harp, he will say, "Yes: it is sweet and tender, like the grass of the plains." Let us renounce therefore the thought of determining the kind of senses with which Nature has endowed the beings who live in ether; for they relate to objects and ideas of which all knowledge is forbidden us.

The story is familiar of the man born blind, operated

upon by the surgeon Childesen, who, having obtained his sight, spent a considerable time in learning how to use his eyes: he had to educate his organs by degrees, and train his reason step by step. Equally familiar is the beautiful fiction of Condillac, about the man whom he imagined to have come into the world lacking the senses of sight, touch, and hearing, and therefore destitute of ideas. In investing him little by little with each sense, the philosopher composes piece by piece a soul that feels and a mind that thinks. This philosophic idea has been much admired. Like the man-statue of Condillac, we are here below nothing but unfinished statues possessed of very few senses; but, when we have reached those upper regions that will open to our grander destinies, we shall be endowed with new senses that our reason foresees and our heart desires.

As was said above, we cannot divine with what new senses the superhuman being may be furnished, because they relate to objects and ideas that we are ignorant of, to forces exclusively belonging to worlds now shut to our inquiring gaze. The kingdom of planetary ether has its geography, its powers, its passions, and its laws. With these objects the new senses of those men exalted in glory will concern themselves. But, as all this is an absolute mystery, it is impossible for us to determine the kind of sense which will enable us to comprehend it, and to enjoy it after our resurrection. What we can only have a presentiment of is the improvement that will be effected in the senses that we now possess, — sight, hearing, feeling, smell, and taste. We may anticipate the improvement of these, in analogy with the extraordinary development of one or the other of them, as witnessed sometimes in animals.

We see the sense of smell intensified in a dog in a degree which confounds our imagination. How shall we explain this fact, well known though it is, that a dog catches the scent of a hare or a partridge which quitted the spot hours ago, and is now leagues away? The keenness of sight in the eagle and birds of prey astonishes us in the same degree. These birds, floating in the clouds, espy prey not so big as themselves on the earth beneath, and swoop upon it in an exactly straight line. The perfection of the sense of touch in the bat is very wonderful. When deprived by accident of sight, the bat supplies the lack so effectually by the touch of his membranous wings, that he can steer himself in the air, and find his way within houses exactly as well as if he could see clearly. How exquisitely sensitive is the hearing of the savage Indian, who, with his ear on the ground, hears the footsteps of an enemy walking a league away; or of the musician who, by practice and natural capacity, is able to detect in an orchestral performance, amid the din of fifty instruments sounding at once, a difference of a quarter-tone in one of them!

Suppose that in superhuman beings these senses have acquired that degree of extraordinary activity which they can attain sometimes in men and animals, and you may form an idea of the power and compass of such a sensorial key-board.

We can also conceive the degree of perfection attainable by the senses of an exalted man, by considering the gain of power that our senses may make through the co-operation of science and art. Indeed, before the invention of the microscope, it was not surmised that the eye could penetrate that world in miniature, well called the "infinitely littles," then absolutely unknown to us: it could not be imagined that in a drop of water, for instance, myriads of living creatures might be seen. These creatures have existed always; but it is only within two centuries that man has been able to see them. Up to that time we were ignorant of our ability to see these microscopic beings. To-day a student of the most limited intelligence looks with indifference upon what Aristotle, Hippocrates, Pliny, and Galen never were able to see or even to suspect. It is the same with the invention of the telescope. The discovery of this instrument, made in the time of Kepler and Galileo, pushed back in an instant the boundaries of human intelligence, and opened to it a domain heretofore closed to its scrutiny. Where Hipparchus and Ptolemy saw nothing, Galileo, Huygens, and Kepler, aided by the telescope, made in one night discoveries that never would have been dreamed of but for this marvellous instrument. The satellites of Jupiter and Saturn; a host of new stars; the phases of Venus; and, later, the discovery of new planets, visible only through the telescope (for Uranus was not known by the ancients); the detection of the spots on the Sun; the resolution of nebulæ into a mass of stars, — all these were the almost immediate result of the discovery of the telescope. Thus it was learned that the human eye, aided by art, could penetrate the farthest regions of the heavens.

Suppose now that vision combined telescopic with microscopic power, — that is, that it could, beyond objects at an ordinary distance, distinguish other microscopic objects, and also the heavenly bodies, invisible to the naked eye, — and you can conceive what the vision of the superhumans must be.

We need not say in what extraordinary proportions the number of our senses would increase, if the eye possessed this prodigious degree of adaptability, — if it could exercise at once telescopic and microscopic functions. Science would take gigantic strides. What advances would not chemistry make, if our eyes could penetrate to the interior of bodies, see their molecules laid bare, and judge of their relative size and arrangement, and the form and color of their atoms. A glance of the eye would reveal to us, as to the inner nature of chemical combinations, what the genius of Lavoisier has not been able to divine. Nature would possess no more secrets for us; for we should learn by simple sight what we now seek painfully by reasoning and experiments, always difficult and uncertain. We should *see* why and how bodies become heated or electrified. We should *see* what light and heat really are. We should have explained to us the mathematical laws under which operate physical forces, — light, caloric, and magnetism. Our unaided eye would be able to solve physical and mechanical problems which have puzzled the minds of Newton, of Malus, of Ampère, and Gay Lussac.

With such a marvellous power we believe the vision of superhuman beings is endowed.

We could apply the same reasoning to the other senses; but it is sufficient to have made plain, by this single example, how those senses which in us are imperfect or rudimentary can be improved and made keener.

We will only add that, as a consequence of the same degree of perfection, these senses must operate with a celerity comparable only with that of electricity or light; that is, that these senses, like heat and light, can operate at a great distance and instantaneously. If the whole body of a superhuman can transport itself with unequalled speed from place to place, as we contend, his senses also can perceive at a great distance. We can hardly err, generally speaking, in comparing with electricity and light the operations of the senses in that invisible world which we dare to traverse.

Is there any sex in superhuman beings? Assuredly not; and in the Christian religion the same question touching the angels has received the same answer. The Christian angel has sometimes male and sometimes female features, — the sweet face of a young man or the tender contour of a girl. Sex is suppressed: the individual is androgynous, a man-woman. So must it be with the superhumans. The reciprocal affection which pervades the dwellers in ether requires no difference of sex.

It should be noted that the affections become spiritualized in the ascent from animals to man. Animals know little of the sentiment of friendship. Love, with its material impulses, holds undivided dominion over them. The affections of animals, beyond that of carnal passion, are limited to the maternal instinct, which is strong and unselfish, but not lasting. The young is the recipient of careful attention, so long as his feebleness demands it; but when he is a little older, and able to shift for himself, he is abandoned by his mother, who no longer recognizes him. Maternal affection, then, has but a short duration in animals. No sentiment endures in them but the love which proceeds from sexual propensities. In man the sentiments of affection are many, often noble and pure. We love child and mother so long as our hearts beat. We love our brothers, our sisters, our relatives, with a tenderness that has nothing carnal in it, and which touches the very roots of our souls. If love is generally indissolubly associated with physical desires, it can sometimes cast them off: a disinterested friendship survives the extinction of the senses. In this respect we are infinitely above animals. Go a step farther, even to the superhuman, the next natural link to our own kind, and we must find the sentiment of affection entirely distinct from sexual difference. In this final blissful home, they have all the same organic type. They need not, in order to love each other, belong to opposite sexes, to two groups of different organizations: their love is the result of the serenity and ineffable goodness of soul, of the sympathy excited by mutual perfection.

In another view, the ethereal home which awaits us

is the place of reunion of those who loved each other on earth. There the son will find the father who was removed from his tender cares; there the mother will see again the adored daughter whom Death snatched from her arms; there husband and wife meet again, and friends joyfully greet each other. But in the new forms that they have taken on, in the improved body that is now the residence of their regenerated souls, there is no more sex. Love is an ennobled sentiment, ideal and exquisitely pure.

How blind and selfish love is here on earth! How narrow and egotistical is our friendship! How difficult it is to spread and extend it to embrace all humanity! Why is it so hard for it to lift itself to the sublime Creator of the Universe? Why do we not love God as we love our neighbors? It will be very different in the upper spheres. Our power of loving, hampered here by the bonds of flesh, will there be freed from all sensual taint. Carnal desire will no longer be the enforced attendant of Love, which will be liberated from all physical temptations and purified of all alloy. Man exalted to glory will love her who was his wife as he now loves his son, his brother, and his friends. The senses will no longer degrade the affections. The happiness that he will feel in the sentiment, purified and strengthened from new sources, will fill and overflow his soul. His capacity for loving will reach all Nature. It will unfold itself into the highest spheres. It will rise to sublime sensations, which will inspire him with this universal love, this great sympathy with all creation. True charity, charity for the whole world, will enfold the heart. The love of God will govern the affections, multiplied by all the height of his infinite power; and the deep glow of tenderness for our fellow-creatures will culminate in sublime adoration for the All-Creator.

But, it will be asked, if there is no sex among superhumans, how is reproduction possible? how do they maintain themselves? how is their kind multiplied? There is no need of reproduction, no need of self-maintenance or self-multiplication of superhumans. The reproduction and conservation of species are the work of the inhabitants of inferior worlds, the Earth and the planets. Such is their lot, and such the obligation that Nature has put upon them. But reproduction is useless and unknown to the dwellers in superior worlds, the blest beings who live in planetary ether. They receive from Earth and the other planets the procession of their new hosts. The battalions of the elect are recruited by arrivals from the lower worlds. Below is the multiplication of individuals: above is the home of the blest; and these have no need of conserving their kind, because the laws of their destiny are unlike those that control the terrestrial or planetary man. Reproduction is the lot of the lower worlds; permanence the privilege of the higher.

The Soul of the Superhuman Being.—In some admirable papers on popular science by Dr. Pouchet, director of the Museum of Natural History at Rouen, and published in the "Univers," we found an idea which forcibly impressed us. M. Pouchet says that Bremser, a German naturalist, lays down the proposition that in man matter and spirit are in almost equal proportions; that is, that he is half spirit and half matter. Bremser bases his theory on the fact that in man sometimes spirit rules and controls matter, and sometimes matter gives laws to the spirit, — all with nearly equal power and completeness on both sides.*

Admitting, with the German naturalist, this proportion to be true, we will say that, while in man the proportion of the soul is 50 per cent, in the superhuman it is undoubtedly 80 to 85 per cent.

---

* "We must consider that man is not a spirit, but only a spirit confined by matter of different kinds. In a word, man is not a God; but, in spite of the bodily captivity of his spirit, it has now become so free in him that he can perceive that he is governed by a spirit higher than his own; that is, by a God.

"It must be presumed that, if a new creation were possible, beings would be born far more perfect than those of preceding creations. In man spirit is to matter in the proportion of 50 to 50, with slight differences on either side; for sometimes spirit, sometimes matter, holds sway. In a subsequent creation, if that which made man is not the final one, there would evidently be organizations in which the spirit would act more freely, and in which it would be in the proportion of 75 to 25.

"Hence it follows that man, as such, was formed in the most passive epoch of our world. He is a sad middle term between the animal and the angel: he reaches after high knowledge, and cannot grasp it; although some of our modern philosophers think they do, they really do not attain it. Man longs to fathom the first cause of all things; but he cannot do it. With inferior mental powers, he would not presume to try to understand those causes which, on the contrary, would be clear to him if he were endowed with a broader intelligence." — *L'Univers*, pp. 760–61. Paris, 1868

It is of course understood that this estimate is advanced only in order to make clear our idea. The single object of our figures is to prove that matters of the mind can be weighed, measured, and compared; a thing generally believed to be impossible.

The soul, then, preponderates in the superhuman being; a point which it is important to know. Let us now try to analyze the soul, as we have analyzed the senses, of the superhuman.

If his senses are many and exquisite, the faculties of his soul, which are intimately connected with the operation of his senses, and depend on their perfection, must be especially active and powerful. We know, moreover, that the faculties of man's soul are feeble and limited. Man lives so short a time on earth that the most potent faculties would be of little use to him, or would not have time to develop themselves or to be profitably employed. But, in the superior world that awaits us, every thing is enlarged and exalted; so that the faculties of the thinking being who dwells in the upper realms must be numerous and of vast compass.

We will say of the soul-faculties of the superhuman what we said just now of his senses. His soul must be furnished with new powers, and those it had on earth must be greatly expanded and improved.

It would be impossible to determine the nature of the new faculties which belong to the superhuman, because they relate to the upper world, which is unknown to us: they respond to moral needs of which we have no idea. Let us not attempt, therefore, to discover their nature, but content ourselves with examining the degree of perfection attainable by those faculties of the soul which are also possessed by man.

Attention, thought, reason, will, judgment, which make us what we are, must acquire in the superhuman peculiar force and accuracy. La Bruyère said that there was nothing rarer in this world than the spirit of discernment. That means that good sense and judgment are very uncommon. One needs to live only a little among men to discover how well founded is this remark. It may be stated, not in misanthropy, that, of one hundred men taken at random, there are scarcely one or two who have good judgment. In most men ignorance, prejudice, passion, oppose the judgment: hence good sense, as La Bruyère says, is rarer than pearls and diamonds. This grand and precious faculty of judgment, of which most men are destitute, is not lacking in the inhabitants of the other world: there it must be the universal rule; here it is the exception.

Of all the faculties, the most valuable for forming broad ideas and comparisons, from which knowledge grows, is memory. But how imperfect, unreliable, and, as we may say, valetudinarian is memory in man! It is absolutely voiceless in the time previous to our birth, and in which we yet existed. We have no recollection of the care bestowed upon us in our earliest years. If our mother died in our infancy, we never knew a mother, she never lived for us. If witnesses who saw us in the cradle did not tell us of what happened at that time, we should know nothing of it: we must see infants in swaddling-clothes, head-bandages, and perambulators, before we can believe that we ourselves went through all that. Darkness covers our first steps, and our cradle, as well as our residence in the maternal bosom; and we see no more clearly before our entrance upon life than we do beyond death.

Memory, which is scarcely developed in man before the second year of his life, and which is extinguished in old age, is subject, even at the highest point of its activity, to a thousand falterings occasioned by sickness or lack of exercise: hence this faculty in us is always very precarious. Undoubtedly it will attain, in the other life, the power, the certainty, and the compass which it wants here below.

Moreover, our memory will be enriched with an incalculable number of new facts. Thanks to the sight and the knowledge of the worlds which surround it, the soul will be able to fix in its memory the geography of a multitude of different places. It will know the physical revolutions, the populations, and the legislation of these thousand countries. The superhuman will know what is on the planets, and their satellites that will pass before his gaze, or which he purposes to visit. Indeed, just as we purpose to visit, in pursuit of knowledge, America or Australia, he will visit and travel through Mars or Venus; and his memory will supply itself also with millions of facts which it will know how to retain, and summon when needed. What a power will a memory so stored, and always responsive, bestow on the mind and the reason!

Languages are only the expression, the union of ideas; and Condorcet has said that science amounts to a well-constructed language. Mathematical science has a perfect language, because the science itself is perfect. The language spoken in the planetary spaces must be perfect, because it expresses all the knowledge that belongs to superhumans; and this knowledge is prodigious. The more the mind knows, the better its expression: the superhuman, being very learned, must have a very expressive language.

It will be also a universal language. That of mathematics is understood by all the people of the two hemispheres: algebra can be read by a Frenchman and a German, or by an Australian and a Chinese, by reason of the simplicity and the perfection of the conventional signs that it employs. This language, which is truly universal, enables us to comprehend that that which is spoken in the planetary spaces may be equally universal, and may serve for all the hosts, without exception, that people the worlds of ether.

Owing to the immense expansion of their faculties, and to the perfection of their language, which is of itself a means of extending and making surer their knowledge, superhumans have a strength of reason and an accuracy of judgment, that, joined to the great number of facts that fill their memories, put them in possession of absolute science. Abstruse problems, before which the mortal mind pauses powerless, and which plunge it into absurdity if it persists in assailing them, — such as speculations about Infinity, the First Cause of the Universe, the essence of Divinity, — all these problems, forbidden to our intelligence, offer no difficulties to these mighty thinkers. What humanity esteem geniuses of the first order — an Aristotle, a Kepler, a Newton, a Raphael, a Shakspeare, a Molière, a Mozart, a Laplace, a Lavoisier, a Victor Hugo — would be among them but feeble-minded folk. No science, no moral idea, is too high for their conception. They see rolling at their feet the Earth and the glorious company of her sisters, the planets; they see the worlds of our solar system gravitating in order and harmony around the grand central Sun, which floods them with his rays; they look from their empyrean heights upon the infinitely various spectacles of elemental wars on our poor globe and its fellows: and more fortunate than the humanity of Earth, of Mars, of Jupiter, and of Saturn, while they admire these works of God, they know the secrets of their course and the causes of their motions. In its moral relations they have penetrated the great Why. They know why man exists, and why they themselves exist. They know whence they came, and whither they are going, which we, alas! know not. Where we can see only confusion, they find order and harmony. To them the designs of Providence are clear; and the events in the lives of men and nations, which we charge upon the injustice or cruelty of God, they see to be just and useful, and deserving of our gratitude.

We believe, further, that in the domain of ether, the home of superhumans, time is an element that counts for nothing. We believe this, because time is not to

God, and because superhumans are brought near by their moral perfections to all spiritual nature, and consequently to God. Yet what confirms us in this belief is the fact that profound grief resists the assuaging touch of time: there is no question of duration in the severest afflictions of the human soul; the loss of a dear one is as poignantly grievous at the end of years as when the blow fell.

Thus time, which is every thing to man, — which is not only, according to the English adage, money ("time is money"), but is also our means of knowledge, our studies and investigations, which are more precious than money, — time counts for nothing in the life of the superhuman. He waits, patient and serene, for the coming to his peaceful home of those whom he loved and left on Earth; and, when the happy reunion is effected, he enjoys with them a happiness with which no anxious thought of the future ever mingles. Thanks to this ignorement, this elimination of time, the superhuman contemplates with unchanging tranquillity, majesty, and serenity the spectacle ever-new and ever-wonderful, the procession of the stars and the grand movements of the Universe.

The Life of the Superhuman Being. — In concluding this inquiry into the attributes of the superhuman being, it remains for us to speak of the life that animates him and gives him the qualities of an active creature.

We have said that, according to our system, the superhuman proceeds from the soul of a man, terrestrial or planetary, who has come to be incarnated in the depths of ether. Is this body destined to perish, sooner or later, to dissolve and surrender its elements again to matter, as happens to the earthly body? Does life withdraw itself from the superhuman's body, and does the soul escape from it?

We think this must be so. Everywhere life implies death, which is its fated limit. We cannot cast anchor in the stream of life. If the soul of the superhuman resides in a living body, that body must die, and its material elements must return to the common reservoir of matter. The torch of life goes out in the upper worlds as on the surface of the planets.

The superhuman, then, in our opinion is mortal. After a time whose bounds we will not essay to fix, he dies; and the soul that was imprisoned in him escapes, as a sweet perfume escapes from a broken vase. What becomes then of the soul, freed from the body on which death has laid his icy hand? That is what we are going to inquire about in the next chapter.

CHAPTER X.

What becomes of the Superhuman after Death. Deaths, Resurrections, and New Incarnations in the Ethereal Regions.

IN animated Nature around us there is a constant scale of gradual improvements, from the plant up to the man. We pass from algæ and mosses, which represent a rudimentary state of vegetable organization, through the whole series of improvements in the vegetable kingdom, and arrive at the inferior animals, zoöphytes and mollusks. From them we rise by imperceptible gradations to the superior animals, and at last reach man. All the degrees of this ascending scale are hardly distinguishable, so delicately managed are the transitions and shades of difference; so that there is a chain of intermediate creatures, in positively infinite length, from the algæ up to us. Some would persuade us that between man and God there is no other kind of intermediaries; that this chain of constant progress leaves a vast void between man and his Creator. They would have us believe that all Nature is arranged, from the vegetable up to man, in successive and infinitely numerous gradations; while between man and God there is nothing but a desert, an immeasurable *hiatus!* Plainly this cannot be; and it is only in their ignorance of the phenomena of Nature that religions and philosophy have fallen into this serious error. There must be between man and God, as between the plant and the animal, a multitude of intermediary creatures which bridge the passage from humanity to the Divinity which rules in omnipotence and infinite majesty.

There are such intermediaries, we are sure. We cannot see them; but if we refuse to admit the existence of whatever we cannot see, we shall fall into cruel absurdities.

Let a man of science show to an ignorant person a drop of water from a neighboring pool, saying, "This drop of water, in which you see nothing, is filled with animals and diminutive plants, which live, are born and die, like the animals and plants in our fields." At these words, the other will shrug his shoulders and think the speaker a fool. But let him consent to place his eye to the magnifying-glass of a microscope, and through it examine the contents of the drop, and he will see that the *savant* told the truth; for in that drop of water in which at first he saw nothing, his eye, effectively seconded by science, will discover worlds.

There may be, then, living beings, and very many of them, where not one has ever been seen; and science can, in this direction, open the eyes of the masses.

We should like to be the *savant* just mentioned. Between man and God, the ignorant herd and a blind philosophy see nothing; but substitute for the eyes of the body those of the spirit, — that is, employ reason, analogy, induction, — and these mysteries will be made clear.

We have already, in treating of the superhuman, described one of the intermediary creatures between man and Divinity, and recognized one of the natural beacons placed on the road to limitless space. But the scale cannot stop at the first being; and we are convinced that these living hierarchies continue, very many more of them, before the radiant throne of the Eternal is reached. Moreover, as we have said, the superhuman, according to our belief, is mortal. What becomes of him after death? We must take up, and we do take up here, the thread of our deductions.

We believe that the superhuman having died, at the end of a period whose bounds we cannot determine, his soul, improved by the exercise of the new faculties that it has received, and the new senses with which it has been endowed, enters a new body, equipped with still more numerous and more exquisite senses, and armed with still more powerful faculties, and begins a new life.

We may give the name of archangel, or an arch-human being, who, in the depths of space, comes next to the angel or the superhuman.

The moment of transition from one life to another must, like the hour of our death on Earth, be filled with physical and mental anguish. Those last minutes, when the metamorphosis takes place in the sentient being, are crises full of agony and torture.

We will not seek to fathom the secret of the organization of this new being whose existence we take for granted, and who is higher in the moral hierarchy than the superhuman, for facilities for investigation would fail us at this point. We brought ourselves to hazard some conjectures as to the body, the soul, and the life of the superhuman, because at this adventurous point in our journey toward unknown spheres we had a term of comparison and induction; that is, the human species. But as to the arch-human next beyond the superhuman, the archangel next beyond the angel, the means of induction fail, because the superhuman himself has been viewed only by conjectures and analogy, which we are unwilling to push farther. We shall cease therefore to prosecute this kind of investigation: we will leave the reader to exercise his own imagination as to the form of the body, the number and perfection of the senses, and the compass of the faculties of that blessed being who ranks next beyond the superhuman, and lives like him in the vastness of ethereal space.

We will only add that, in our belief, not at the second, nor the third, nor the fourth incarnation can stop the chain of sublime creatures that we have caught a glimpse of, floating in the celestial infinite, and which spring from a soul first human, and successively enlarged and exalted in improvements and moral power. The number of those beings, advancing ever nearer toward perfection, that follow one after another, it seems to us impossible to estimate, with no lights but those of reason and acquired knowledge. We can only say that the creatures who compose this chain of perfection must be very numerous.

At each elevation in the hierarchy of Nature, the celestial being sees expand the wings which give us an idea of his wonderful power. Each time his organs become more numerous, more flexible, more comprehensive. He acquires new and exquisite senses. He has greater and still greater means of extending everywhere his beneficent empire, of exercising his faculties, of loving his kind and all Nature, and especially of comprehending and blessing the Divine intentions. A tender affection, deeper and still deeper, possesses his soul; for affection, and the happiness which it engenders in his profound content, are given him as consolations for the anguish of death to which he is inevitably doomed.

Thus the bliss of the elect is enhanced. Thus the beings who dwell in the boundless plains of the invisible world employ each of their lives in preparing for the life which follows it, and in insuring to themselves — by the judicious use of their liberty, by the cultivation of their faculties, by the conservation of their moral health, by continual benevolence — a life more noble, more animated, and happier, in the new realms that open to their lofty destinies.

Nevertheless, as everything comes to an end on this earth, everything must have an end on the spheres that surround it. Having passed through this long succession of halting-places and heavenly stations, these beings must finally arrive in one place. What is that place, — the final limit of their vast journey through space? As we believe, it is the Sun.

CHAPTER XI.

Physical and Geographical Description of the Sun.

THE Sun, according to our theory, is the centre at which are reunited the souls come last from ether, and originally from the several planets. Having undergone in the ethereal plains the successive incarnations that we have described, these souls, originally human, at last reach the Sun, and pass into the domain of the King-Star.

This, then, is the place to describe the Sun, from a physical and astronomical point of view. This description will help us to comprehend the truly sovereign office of this peerless globe. The astonishing attributes that belong to it alone, the inconceivable power that it wields, will sufficiently show why we place the Sun at the very summit of the ladder of Nature.

First, the Sun differs utterly from the other stars of our world. It is like nothing, and nothing can be compared to it. Neither the planets nor the satellites, nor the asteroids nor the comets, can give us an idea of it. Its immense size, its physical constitution, its incomparable properties, give it an isolated rank, and will always warrant its assignment to a place by itself and sovereign.

The enormous size of the Sun helps us to understand its supremacy. It is large enough to furnish a home for all that can be sent to it from all the planets. Its own volume is larger than the united volumes of all the other heavenly bodies that revolve around it. It is six hundred times larger than all the other planets, together with their satellites, and the asteroids and comets that compose what is called the solar world; that is, the world of which we are a part.

Since it is larger than all the other stars put together, the Sun must be greater than the Earth. But in what proportion does it surpass the Earth in size? It is *one million three hundred thousand times larger* than our globe.

Illustration alone can give a fair idea of the comparative size of the Sun and the several planets. For this reason we place before the reader a figure (Fig. 14), which shows exactly the proportionate magnitude of the Sun and the largest planets of our world. The Earth, represented by a point, indicates what must be the size of Mars, Mercury, and Venus, which are still smaller than the Earth.

Some simple comparisons will aid us in representing effectively these proportions.

If we imagine the Sun to be a sphere a decimetre (3.937 inches) in diameter, the Earth must be a millimetre (.03937 inches) in size.

The terrestrial globes used for teaching geography in our schools are usually thirty centimetres (0.39371 inch) in diameter: a solar globe, which should present an exact proportion to the terrestrial globe, would have to be thirty-three metres (1.093633 yards) in diameter.

If we imagine a spherical balloon big enough to reach half-way up the tower of Notre Dame, while resting on the ground, and make it represent the Sun, a globe three decimetres in diameter would represent the Earth; one of three metres and thirty centimetres, Jupiter; and one of three metres, Saturn.

It takes three years to sail around the world. To make in the same manner a tour of the solar globe would occupy three hundred years. If longevity were not greater in the Sun than on the Earth, a lifetime

Fig. 14. — Comparative magnitudes of the Sun and the planets.

would not be long enough for a traveller there to see the whole surface of the globe he lives on.

The force of gravity is nearly thirty times greater on the surface of the Sun than it is on the Earth. We know that a body falling on the Earth traverses in the first second of its fall a distance of four metres, nine centimetres. In the Sun a body traverses one hundred and forty-four metres in the same time. Hence one of our bodies, transported to the Sun, would weigh about 2,000 kilogrammes (about 4,500 pounds), — as much as an elephant's. The body of a dog or of a horse would weigh twenty-eight times as much as on our globe. The natural conditions of the Sun must be far different from those which exist in the group of planets of which the Earth is a part.

The Sun emits continual fires; a peculiarity which is not shared by any other stars of our world. It burns of itself, and scatters heat and light abroad. On the contrary, the other stars of our world are neither hot nor luminous; and, if the Sun were not, they would be plunged in eternal darkness, doomed to everlasting cold. This is a peculiarity which aids us to understand the fundamental part assigned to the central star.

The light and heat from the Sun are steady: they are never intermitted, never lose their power. This second characteristic — the constancy of illumination — still more strikingly distinguishes the Sun from the other celestial bodies of our world.

The intensity of the Sun's actual heat has been

measured by philosophers. They have reached this end by ascertaining by experiment the quantity of heat accumulated in a given time on a specified spot of the Earth exposed to the Sun, and adding to this amount the quantity which must have been absorbed by the atmospheric air, by the ethereal spaces, and by the soil.

The French philosopher Pouillet, who has given much attention to this delicate investigation, reaches conclusions which he thus states: —

> "If the total quantity of heat emitted by the Sun were exclusively employed in melting a bed of ice, which completely surrounded the solar globe, this quantity would suffice to melt in one minute a bed of ice eleven metres $\frac{80}{100}$ [about 32 feet], or in one day seventeen kilometres [nearly 50,000 feet]."

"This same quantity of heat," says the English *savant* Tyndall, "would in an hour set boiling 2,900 milliards of cubic kilometres of ice-cold water." *

The astronomer, John Herschel, discovered that in order to extinguish the Sun, and prevent it from "radiating caloric," as the scientific term is, one would have to project upon its surface a jet of frozen water, or a cylindrical column of ice, eighteen leagues in diameter, and propelled at a speed of 70,000 leagues per minute. This proposition Herschel expresses as follows: —

* These figures are too prodigious to be expressed in English. Remember that a milliard is 1,000,000,000, that a metre is more than three feet, and that a kilometre is equal to 1,000 metres. The reader may multiply for himself, and write down the result — if he has paper enough. — Tr.

"Imagine a cylindrical column of ice, eighteen leagues in diameter, incessantly dashed upon the Sun and the melted water instantly removed. In order that the entire solar heat should be employed in melting the ice, without any external radiation, this frozen cylinder would have to be driven upon the Sun with the speed of light. Or, in other words, the heat of the Sun could, without diminishing its intensity, melt in a single second a column of ice 4,120 kilometres square [4,503,160 yards] at the base, and 310,000 kilometres [338,830,000 yards] high."

A comparison made by Tyndall still more clearly illustrates the intensity of the Sun's heating-power. "Imagine," he says, "the Sun to be enveloped in a coating of coal nearly seven leagues (27 kilometres) thick. The heat produced by the combustion of this coal is what the Sun yields in an hour."

Philosophers have ascertained exactly the intensity of the Sun's light, as well as that of its heat.

It has been proved that the solar light is three hundred thousand times stronger than that of the full moon, and seven hundred and sixty-five million times stronger than that of Sirius, the most brilliant of the stars.

Bouguer found, by experiments made in 1725, that the Sun, at the height of 31 degrees above the horizon, had an illuminating power equal to that of 11,664 candles placed forty-three centimetres from the object illuminated, or to 62,177 candles placed one metre distant. According to this discovery, — account being taken of atmospheric absorption, and the law of the variation in the intensity of light, which diminishes in inverse ratio to the square of the distance, — the illuminating power of the Sun at the zenith would be seventy-five thousand two hundred times as great as that of a candle placed one metre from the object.

The English philosopher, Wollaston, reached a like result. By experiments conducted in another manner in the months of May and June, 1799, Wollaston found that 59,882 candles, at the distance of one metre, gave as much light as the Sun. Supposing the Sun to be at the zenith, the illuminating power, according to this estimate, would be equal to that of 68,009 candles.

It will be seen that this estimate differs but slightly from that of Bouguer, who reached the figure of 75,200 candles.

Intense as is the light of the Sun, we now possess sources of illumination which approximate it. Such is the oxyhydrogen light, which is made by burning hydrogen gas in a current of oxygen gas or air; a mode of lighting which has been recently employed in industrial establishments in Paris and London. This light possesses an illuminating power of more than 200 candles. A thread of magnesium burning in the air develops an immense quantity of light, which has been estimated as equal to 500 candles. In fine, the electric light yielded by a voltaic pile of sixty or eighty couples produces a luminous bow as brilliant as five hundred, or eight hundred, or even a thousand candles. In the last case, the voltaic bow, regarding it according to the estimates of Bouguer and Wollaston, would emit seventy-five times less light than the Sun, supposing the luminous electric point to be at the distance of one metre.

With very powerful piles even more has been accomplished, and light has been obtained not much inferior to that of the Sun. MM. Fizeau and Foucault, comparing the brilliancy of a voltaic bow produced by the action of three series of Bunsen piles, of forty-six couples each, with the light which the Sun gives in a clear sky in the month of April, have determined that the illuminating power of the Sun is not more than two and a half times greater than that of the electric light.

The foregoing figures represent the illuminating power of the Sun, viewed with reference to our globe, atmospheric absorption being taken into account. Arago, trying to ascertain the intrinsic illuminating power of the Sun, found that the intensity of solar light is thirty-two thousand times greater than that of a candle shining at the distance of one metre. According to later researches made by M. Becquerel, the result obtained by Arago would be far below the truth, and the brilliancy of the solar light would be one hundred and eighty thousand times greater than that of a candle shining at the distance of one metre.

In speaking of the several planets, we have mentioned the distances which separate each of them from the Sun. We need not recur to this point. But what should be carefully noted in order to show the altogether sovereign part played by the Sun in the economy of the world is the fact that all these planets, escorted by their satellites, as well as the comets which casually appear to us, revolve around the Sun. The

Sun stands motionless in the midst of this majestic procession of stars, which move around him like so many courtiers rendering homage.

Thus the Sun is, as it were, the heart of our planetary system: all concurs with, all converges towards, him.

Some persons, only half-informed in science, will cry, "That is very simple! The Sun being six hundred times larger by itself than all the other stars taken together, the phenomenon of the revolution of all these stars around it is explained by the law of attraction, which declares that bodies attract each other in direct ratio of their masses. If the Sun attracts to itself the stars of our world, it is because its mass is greater than that of them all together." He who makes such an answer is guilty of the common error of mistaking a word for a thing, a hypothesis for an explanation, in putting a term of language in place of a reason. When Newton imagined the hypothesis and the phrase of the reciprocal attraction of matter, he was very careful to say that he proposed only to give a name to a phenomenon wholly inexplicable in itself, and of which we know only the external manifestation; that is to say, the mathematical law. We know that bodies approach each other in the ratio of their masses, and in inverse ratio of the square of their distances; but why do they approach each other? This is what we do not know, and what we probably never shall know. The word "attraction" has superseded, since Newton's time, the word *tourbillons* (vortex) that Descartes had brought into use in his own day. If for the term "attraction" we substitute the term "electrization," or indeed, as Kepler did, the terms "affection," "sympathy," "obedience," &c., we should have a new hypothesis, with a new term; and the mathematical law of the manifestation of this electrization, affection, sympathy, or obedience, would be always the same, only the hypothesis would be changed. As to the real cause which makes small bodies to rush towards greater ones, and the little stars to revolve around the larger, it is, we repeat, a mystery that cannot be penetrated by mortals.

Whatever hypothesis may be taken to explain this fact, it is certain that the Sun holds suspended above the depths of space the planets with their satellites,— the asteroids, and the comets which journey through the heavens, never passing beyond his directing influence. The Sun draws after him all the stars, which follow and surround him, like flatterers of his power, humble vassals of his universal pre-eminence. Like the father of a family in the midst of his progeny, the Sun holds peaceful sway over the many children of the sidereal creation. Obedient to the irresistible impulse that emanates from the central star, the Earth and the other planets circulate, revolve, and gravitate around him, blessed with his beneficent floods of light, heat, and electricity, which are the prime agents of life. It is the Sun that marks out for the planets their courses in the heavens, while at the same time he proportions their days and nights, their seasons and climates.

The Sun, then, is the hand which sustains the stars above the unfathomable abysses of space, the central fire which warms and the torch which lights them, and the source from which they draw the principle of life.

The grand and truly unique office of the Sun in the economy of Nature has always been understood; but not until our day was this truth established by profound investigations. Science has even transcended what the imagination of poets has conceived as to the preponderance of the Sun in our world. By multiplied experiments and abstruse calculations, modern philosophers have proved that the Sun is the first cause of almost all the phenomena which occur on our globe; and that, without the Sun, the Earth, and no doubt the other planets, would be nothing but vast deserts, gigantic corpses, so to speak, revolving, useless and frozen, in the wastes of infinite space.

It was the English philosopher Tyndall, who, comparing many new facts in natural philosophy and mechanics, has demonstrated this truth; and it may be said that the results to which this investigator has been led constitute what is, perhaps, the most brilliant page of contemporary physics.

We will try to explain how every thing on the Earth, and doubtless on the other planets also, springs from the Sun; so universally, indeed, that we may affirm that vegetables, animals, man, and the whole living creation, are but the offspring, the children, of the Sun, — that they are, as it were, woven of solar rays.

And, first, the Sun is the prime cause of all the movements, great and little, that we witness in the air, the waters, and the soil; and which nourish on the surface of our globe the activity, feeling, and life.

Take the winds, for example, which have so important a bearing on all the physical phenomena on this globe. Whence come they? From the action of the Sun. In fact, the Sun heats different parts of the Earth very unequally: he scorches the tropical and equatorial regions, and leaves other latitudes colder. From another point of view, on every spot of the Earth touched by the Sun's rays, *strata* of air near the ground expand and rise: they are instantly replaced by colder *strata*, come from temperate regions. In this way arise those periodical winds, called "the Trades." For this reason, in each terrestrial hemisphere two great aerial currents blow constantly, going from the equator to either pole: the upper one tending toward the north-east in the northern, and toward the south-east in the southern hemisphere; the lower, which holds a contrary course, going from the north-east or south-east.

The motion of the Earth causes other regular winds. The action of heat and evaporation, combined with the unequal distribution of continents and water, produces others, which are irregular. Thus, in the great valleys of the Alps and Cordilleras, the heating of the air determines the afflux of cold air from the mountains, and brings about violent winds, — real whirlwinds.

Sea-breezes are occasioned by differences of temperature on the shore in the day and night. During the day, the Sun has heated the shore, and caused a considerable expansion of the air. When the Sun sets, the warm air is replaced by fresh currents coming from inland. The same phenomenon is repeated in the morning, on the return of the Sun: the shore being heated by his rays, the warm air rises, and is replaced by the colder air of the sea which tends toward the land. Therefore the evening breezes blow from the shore, and the morning breezes from the sea.

To the successive appearance and disappearance of the Sun, then, may be ascribed those mighty agitations of the atmosphere that we call winds, and the less violent ones, called breezes. The positions of the Sun, which vary constantly according to the time of the year and the hour of the day, explain the inequality, yet uninterrupted existence, of those aerial currents.

The heat of the Sun, which dilates the atmospheric air, and the absence of this heat, which contracts it, — these are the general causes of the winds, which serve to keep in the same state of homogeneity the air of all terrestrial regions.

The watering of the globe — that is, the rain, an element indispensable to the exercise of the functions of life — is also a consequence of solar heat. The waters of the sea, the rivers, streams, and lakes, those which impregnate the soil, or are exhaled from vegetable masses, are gradually vaporized, and form clouds or invisible vapor. When the Sun sets, these vapors are cooled in the bosom of the atmosphere in which they float, and fall back to Earth in the shape of dew, fog, and rain.

If the cooling of the vapor of water in the atmosphere is more violent, instead of rain, we have snow; that is, a fall of frozen water. Snow falls, and accumulates mainly on the summits of mountains, because the temperature of those lofty points is always cold. In the highest altitudes, snow remaining a long time passes into a peculiar state, intermediate between snow and pure ice, and finally forms those masses of frozen water that we call glaciers. In warm seasons, glaciers melt gradually: the resulting water flows down the roofs of the mountains into the valleys, where it becomes the source of springs, rivers, and streams. These rivers and streams flow toward the ocean, from which they are again evaporated by the action of solar heat.

Thus is effected and maintained the continual exchange, the incessant circulation of the waters, which pass from the surface of the Earth into aerial masses, and which effect the watering of the globe, — a phenomenon essential to the operation of vital functions.

The regular currents established in the depths of the ocean also result from the action of solar heat. From the poles to the equator, the water of the ocean is unequally heated; and this inequality in temperature produces, from the poles to the equator, a regular furrow caused by the displacement of the water, the cold currents hurrying to replace the warm. The unequal evaporation, caused by the unequal distribution of heat at the equator and the poles, helps to the same end, by increasing the saltness at the equator without causing a corresponding increase at the poles, which occasions a difference of density, and finally a displacement through non-equilibrium. The currents of the ocean are therefore, in part, caused by the action of the Sun.

Thus the winds, the irrigation of the globe, and the currents of the sea, are caused by solar heat.

The movements of the magnetic needle are also occasioned by the influence of the Sun, if it be true, as Ampère said, that the magnetic currents which groove the Earth are simply thermo-electric currents, occasioned by the unequal distribution of heat on the surface of the globe.

While the Sun is the agent of powerful physical forces, he is also a valuable chemical agent; and in this capacity the greatness of his presence in the phenomena of Nature is especially conspicuous. The light and heat of the Sun govern, on the surface of the Earth, the most important chemical processes, which are closely connected with the exercise of the functions of animal and vegetable life. If the Sun were not, life would be banished from the globe. Life is the daughter of the Sun: this is what we shall try to show.

The processes of photography will help us to understand how the light of the Sun presides over the chemical operations which are going on in the bosom of vegetable life.

What is photography? What is the curious phenomenon which enables us to fix on paper a design formed by the light? A sheet of paper impregnated with chloride or iodide of silver being placed in the focus of the lens of a camera, on it, moistened with water, is cast the image formed by the lens. Those parts of the image not illuminated produce no effect on the salts of silver, which is incorporated with the paper; while those parts on which the light falls decompose the salts, and turn it violet or black. If you withdraw the paper from the apparatus, working in the dark, you have a design which reproduces in black the luminous image formed by the lens. You can readily fix and make unchangeable this image, which is the simple product of the chemical action of light.

All the salts of silver exposed to the light in this way undergo a like decomposition.

But the salts of silver are not the only substances that are modified by the light. Compounds of gold, platinum, and cobalt, properly prepared, can be similarly changed under the influence of luminous rays, direct or indirect; that is, exposed to the Sun, or to a diffused light.

The light of the Sun has the property of promoting combinations among many other bodies. Such happens in the case of hydrogen and chlorine. Mix these two gases in a bottle, open to the light, equal parts of each, and expose the mixture to the Sun: an instant combination of the two gases is effected, and they become chlorohydric acid gas. The combination is effected with such violence, that considerable heat is thrown off in the process. If a bottle containing the mixture be thrown into the air toward a spot illuminated by the Sun, the bottle, before reaching the ground, bursts, with a tremendous explosion, the instant it passes the limits of illumination.

We could multiply examples of the chemical effect produced by light alone on substances belonging to the mineral kingdom. Without dwelling longer on this point, we will say that the chemical action of light is even more powerful and more general in the vegetable than in the inorganic kingdom. There may be seen a phenomenon so important, that we cannot fail to see in it a veritable premeditated design of Nature.

One of the most fruitful discoveries of modern science is the recognition of the fact, that the respiration of plants takes place only in the presence and by the direct action of light; that is to say, that the decomposition of the carbonic acid, which circulates in the tissues of vegetables, and which has been drawn from the soil by the roots, is effected only where the plants are exposed to the Sun. We know from the works of Priestley, Charles Bonnet, Ingenhouze, and Sennebier, that the decomposition of carbonic acid in carbon, which remains fixed in the tissue of the plant, and in oxygen, which is evolved, can happen only in presence of the Sun's rays, direct or indirect. Every one of our readers can convince himself of this fact. Let him place in a drinking-glass filled with water a handful of green leaves, and expose the whole to the Sun: in a day's time, he will see the upper part of the glass filled with some centilitres * of a gas, which

* A centilitre is equal to 0.338 fluid ounces.

is nothing but pure oxygen proceeding from the respiration of the leaves. We shall fully appreciate the importance and the value of such a phenomenon, if we remember that it happens throughout the whole extent of our globe, and that respiration — that is to say, the life of all the vegetable masses which cover the Earth — depends exclusively upon the light of the Sun. It is by the respiration of plants, which returns oxygen to the atmospheric air, that Nature supplies the deficiency of the same gas in the air, caused by the respiration of animals, and by the continual absorption of this gas by many mineral substances, as well as the numerous combustions, natural or artificial, which are constantly occurring on our globe. All these combustions would result in exhausting most of the oxygen of the air, if there were not a permanent cause of its restitution. This permanent cause is the respiration of plants, induced by the solar light.

It is so true that the respiration of plants is dependent on the action of the light of the Sun that, if the clouds were to intercept that light, the evolution of oxygen by the respiration of plants would undergo a marked diminution. If the light of the Sun ceases abruptly, as in a total solar eclipse, the evolution of oxygen is interrupted, and plants permit the carbonic acid to exhale unmixed, as happens during the night.

It is for this reason that a plant kept in absolute darkness loses its colors, and grows pale: it no longer breathes; it discharges carbonic acid gas without retaining its carbon; it *withers*, according to the common phrase. This means that the plant no longer lives on the outer air, or on the gas furnished by the soil: it consumes its own substance. Lettuce has the pale color that we notice, because it has grown in darkness; and the mushrooms that we eat are so white simply because they are grown and matured in dark caves.

M. Boussingault, who made a study of vegetation in darkness, discovered that the leaves of a vegetable, which sprang up and was developed in a place profoundly dark, never exhale oxygen: its respiration yields only carbonic acid gas. The plant breathes, then, as an animal would breathe. Yet it must be remarked that it is only the substance of the seed that provides for this production. The plant borrows nothing from outside: it consumes only the elements contained in its seed; and, when these sources of nutrition are exhausted, it dies. The duration of its life depends only on the weight of the seed which has germinated.

If it is a well-developed plant that is kept in darkness, the same thing is noticed. The plant evolves nothing but carbonic acid; and, as it borrows nothing from outside, it perishes when it has devoured its own substance.

M. Sachs says, in his "Vegetable Physiology," that the peculiar movements of the leaves of many vegetables cannot be made if the plant is kept in darkness. Shut up in the shade, plants remain always a prey to that state which has been called "sleep," since the time of Linnæus.

The colors of flowers are produced, it is true, on the

interior of natural envelopes, which shield them to a great extent from the action of the light; yet it must be noticed that flowers cannot form on the inside of these envelopes except at the expense of substances contained in the leaves, and that the leaves themselves could not put forth but under the influence of the light. It is the same with fruits.

Leaves, flowers, fruits, are then, as a German physiologist, Moleschott, says, "beings woven of air by light." "When we contemplate," says the same author, "the brilliant colors of flowers, and sweet perfumes engender a serene satisfaction in the poetic soul which sleeps within every man, it is still the light that is the mother of the color and the perfume."

The influence of the Sun on vegetation is, then, of an altogether fundamental importance. Without the Sun there could be no plant on our globe. In regions habitually neglected by this powerful and beneficent torch of Nature, — that is, towards the extreme north, — vegetation is stunted; and, when one goes still higher, there is none. The absence of light and the cold are the causes of this complete failure of the natural decoration, and the useful gifts that vegetation gives the earth. In warm regions vegetation is vigorous and extensive, according as the light of the sun is liberally poured upon it. Nothing is comparable to the luxurious vegetation of tropical countries in both hemispheres. Brazil, the interior of equatorial Africa, those parts of India lying between the tropics, — these are the regions most renowned for the vigor and abundance of vegetable life.

Agriculture, enlightened by modern chemistry, has demonstrated the peculiar importance of the Sun in stimulating the energies of vegetation, and producing such combinations of substances as no other influence could effect. M. Georges Ville, Professor in the Museum of Natural History at Paris, is convinced by his own experiments that the Sun works veritable miracles by the vigor with which he inspires vegetable growth. No chemical fact or theory, says the learned Professor, can explain the mystery of the solar influence and its prodigious effect upon the growth of vegetables, and the product of cultivation.

In leaving this subject, we may remark that, owing to a circumstance which seems providential, successive human generations reap the benefit of the chemical power of the Sun that Nature has garnered in certain vegetables through countless ages. What, indeed, is the coal that cherishes our manufacturing industry, which fires our steam-engines, steam-boats and locomotives? It is the residuum of gigantic forests that covered the globe during the geologic periods. The substance of the woods of the ancient world was first changed into turf. This, in accumulated centuries becoming compact and heavy, at last forms the hard and ponderous substance that we call coal, or carbon of the earth. But what was the cause? what was the first agent that produced those forest trees in antediluvian ages? It was the chemical force of the Sun. This force, or, in other words, the products of the chemical force of the Sun, were accumulated and treasured in the wood, and then in the coal which was formed of the wood. We find it there to-day, and use it for our own benefit.

So the burning Sun which heated the lands of the ancient world is not lost to us. They are the same rays, the same chemical force, that contemporaneous generations inherit. The power of the Sun, having slept in the coal for thousands of years, awakes for us: it comes to life again, and transforms itself in our hands into a mechanical power.

The light and heat of the Sun, so important in the vegetable world, wield a like power in the animal kingdom.

If we remember that plants are indispensable for the nourishment of the majority of animals, that the creation of vegetables necessarily preceded that of animals on the Earth (since the latter live on vegetables), and that animals would surely vanish from the Earth if plants were suppressed, we shall be led to see that animals are as truly, though indirectly, the children of the Sun as plants themselves.

But it can be shown that the action of the Sun is directly and immediately indispensable to the maintenance of animal life.

Is it not true, to begin with, that light and heat have a great influence on the health of man and animals? To understand this, we need only compare men who pass most of their time outdoors and men who live in houses shut from sunlight and daylight in the cellars of narrow streets in great cities. Besides being unwholesome by reason of dampness, these habitations are fatal to health because they lack the vivifying presence of the Sun.

Light, which is absolutely indispensable to the exercise of respiration by plants, is not so essential — indeed, far from it — to the breathing of animals. Yet it is certain that the products of the respiration of man and animals are less abundant in the night than in the day time. Moleschott discovered that the amount of carbonic acid gas exhaled by an animal increases with the intensity of daylight, and that it is at its minimum in absolute darkness; "which means," adds the author, "that the light of the Sun accelerates the molecular movement in animals."

Thus the rays of the Sun constitute a preliminary condition of animal life, whether because they induce the growth of plants, the essential basis of nourishment for men and animals, or because they direct the operation of many of the physiological functions of these latter.

Mr. Tyndall, who, in his work on Heat, has developed views like the foregoing, has summed them up in such beautiful language, that we cannot deny ourselves the pleasure of copying it: —

"As surely as the power that sets the clock in motion proceeds from the hand which has wound it up, so surely does all terrestrial force flow from the Sun. To say nothing of volcanic eruptions, of the flow and reflow of the waters, every mechanical action which takes place on the surface of the Earth, every manifestation of power, organic or inorganic, vital or physical, has its source in the Sun. Its heat keeps the sea liquid, and the atmosphere gaseous; and all the storms which disturb them both are the breathings of his mechanical force. He fixes the springs of streams and glaciers in the sides of mountains; and in consequence cataracts and avalanches hurl themselves down with a vehemence which is his direct gift. Thunder and lightning in their turn declare his power. Every fire that burns, every flame that glows, distribute heat and light that once were the Sun's. In our day we are, alas! compelled to hear of battle-fields: now every charge of cavalry, every shock of opposing battalions, is the use or abuse of the mechanical power of the Sun. The Sun comes to us in the shape of heat: he leaves us in the shape of heat; but between his coming and his going he generates the various forces of our globe. All these are special forms of solar force, so many moulds in which it abides for a moment in its passage from its source to infinity."

Philosophers have succeeded in estimating the mechanical force that represents the heat of the Sun; and the calculations by which this end has been reached are curious to contemplate.

To understand how it is possible to express in units of mechanical force an agent of heat, we must give a summary of a theory which is the most beautiful creation in the natural philosophy of our times: we mean the mechanical theory of heat, or the doctrine of the mutual transformation of physical forces.

Experiments have shown that heat changes under our eyes into a mechanical force. See, under the piston of a steam-engine, the steam cooling, and the vanishing heat producing instantly a mechanical power, — and you will understand how it can be maintained that heat turns into force. This being admitted, it will be explained how one of these elements can be represented by the other; or, at least, how force and heat can be calculated by one and the same unit of measure.

This common unit of force and heat is called a *calorie.* A *calorie* is the quantity of heat required to raise the temperature of a kilogramme (2.2046 pounds) of water one degree.* On the other hand, a kilogrammetre is, in mechanics, the amount of force required to raise in one second the weight of one kilogramme to the height of one metre (39.37 inches).

Philosophers have succeeded in solving the difficult problem, how many kilogrammetres there are in a *calorie*, transmuted into mechanical power. It is known to-day, through the admirable works of MM. Mayer, Joule, Helmholz, Hirn, Regnault, and others, that a *calorie* is equal to 425 kilogrammetres; that is, that the amount of heat required to raise the temperature of a kilogramme (2.2055 pounds, av.) of water one degree, Centigrade, produces mechanical power represented by the elevation of a weight of 425 kilogrammes to the height of a metre in one second. This number of kilogrammes is called the mechanical equivalent of heat.

With these *data*, we can estimate in units of mechanical force the power which the solar heat produces in transforming itself into mechanical force. And if we take the total of the solar heat poured upon the Earth in a given time, we can estimate the sum of the forces which all the heat thereon distributed would develop on the surface of the Earth, if it were all employed to supply mechanical power.

---

* Centigrade.

In one year, each square metre (3.280 feet) of the surface of the Earth receives 2318.157 *calories;* that is, more than 23 milliards (23,000,000,000) of *calories* per hectare (2.471 acres), or 9,852,200,000,000 kilogram-metres per hectare. In order to comprehend the intensity of this force, we must imagine a steam-engine, which, instead of the power of two or three hundred horses, like those of our great steamers, combines a power of 4,163 horses. These figures, it should be understood, apply to only a single hectare of earth. Such is the force that the Sun distributes, in a single year, over a single hectare.

Making the calculation for the entire surface of the globe, we have a total of 217,316,000,000,000 horse-power. To get an idea of such a force, we must picture to ourselves 543,000,000,000 steam-engines, each of 400 horse-power, running day and night without stopping. Of such importance to our globe alone is the heat of the Sun!

Of this enormous force, the physical and mechanical operations on our planet, vegetation, the phenomena of animal life, manufacturing and agricultural operations, absorb only an infinitely small part. Mr. Tyndall, in the work from which we have quoted, says on this point: —

"Consider the aggregate of the forces of our world: the power gathered in our coal mines, winds, rivers, fleets, armies, cannons. What does it amount to? A fraction of the Sun's force, equal at the most to $\frac{1}{2160000000}$ of his total power. Such is, in fact, the proportion of solar force absorbed by our globe; and yet we convert only an infinitesimal fraction of this fraction into mechanical power. If we multiplied all our force by millions of millions, we should fail to represent the total expenditure of the Sun's heat."

We have, in this chapter, analyzed the several physical and vital effects produced on our Earth by the light and heat of the Sun. We have reviewed its action on inanimate as well as animate Nature. We have seen that the Sun is truly the grand cause of physical movements on this globe, and that he is also the first principle of life, vegetable and animal. Without the Sun, life would be banished from the Earth: as we said before, life is the daughter of the Sun.

The words "heat" and "life" are almost synonymous. It is said in all languages that creatures are "cold in death:" we speak of "mortal cold," &c. This image is an exact expression of the reality. An animal, or a plant, deprived of life, is necessarily cold. A chill is the harbinger of every malady, and the forerunner of death. Every dead body is a cold body. We could say that in the animal cold takes the place of life, as in brute bodies cold succeeds heat.

If we reflect now that it is only by the continued action of heat that plants can spring up, grow, and mature; that every plant requires, in order to thrive, a certain number of degrees of heat, and that botanists and agriculturists know exactly the number of degrees needed by cereals in order to ripen, and by fruit trees to bear; if we reflect, on the other hand, that a long and quiet accumulation of heat is indispensable to the vivifying of the fecundated egg of a bird, so that by the mere use of heat in artificial hens the want of a natural hen can be supplied; if we reflect also that the eggs of viviparous animals find this heat in the bosom of the mother, and that, moreover, every living thing, as Harvey said, proceeds from an egg (*omne vivum ex ovo*); if we remember that, after the development of the germ in mammiferous animals, the steady influence of maternal heat is indispensable for forming the organs of the fœtus, — we shall, perhaps, be led, by combining all these observations, to inquire if heat does not directly produce life; does not transform itself into vital power. Modern philosophers, who have created the mechanical theory of heat, — that is, the wonderful and profound doctrine of the mutual conversion of forces, — *savants*, who have shown by mathematical demonstration that heat converts itself into a mechanical force, and conversely, can, perhaps, complete their brilliant synthesis by adding that heat, which converts itself into a mechanical force, can also transform itself into life, or into vital force; and that the magnificent theory of the transformation of forces applies not only to brute bodies, but finds in living bodies a thrilling confirmation.

Thus heat and life would be the manifestation of the same power; and the cause of life, like the cause of mechanic force, would reside in the King Star, — the Sun.

CHAPTER XII.

The Sun the Definitive Home of Souls that have reached the Highest Stage of the Celestial Hierarchy. The Sun the Final and Common Residence of Souls come from Earth and the Planets. Physical Constitution of the Sun. He is a Mass of Burning Gas.

THE fundamental importance of the Sun in the general economy of our world having been established, we shall not be surprised to find ourselves in this radiant and sublime abode of human souls, gathered from the different planets, and successively purified and perfected by a long series of incarnations in the depths of interplanetary space. Some philosophers have caught a glimpse of this truth. We have already said that the astronomer, Bode, established the loftiest intelligences in the Sun. "The happy beings who inhabit this blest abode," says Bode, "need not the alternate succession of night and day: a light pure and inextinguishable shines always in their eyes. Amid the glory of the Sun they enjoy sweet security in the shadow of the All-powerful."

In what form must we depict these dwellers in the Sun? We cannot answer this question without an acquaintance with what is called the geography of the Sun, or, as astronomers term it, his physical constitution. In this latter respect the Sun differs essentially from the planets and their satellites, as well as from the comets. Rendered absolutely unique by his place in the Universe, he must possess a peculiar constitution. What, then, is it? What is the geography of the Sun?

We wish we were able to answer this question with nice precision: we would like to describe the configuration of the Sun as confidently as we have described that of the planets. Unfortunately Science has not yet advanced so far. The problem of the real nature of the Sun is full of uncertainties. Astronomers hesitate between two opposing theories, and the one which seems to be the most logical is of too recent a date to be expounded dogmatically. We can only make known the actual state of scientific knowledge in this question, explain that theory which seems conformable to known facts, and apply it to the object on which we are intent; that is, try to discover from it the form, which, as we believe, must belong to the inhabitants of the radiant star.

Up to the date of the discovery of the telescope, — the beginning of the sixteenth century, — in the days of Kepler and Galileo, there had been only vague and arbitrary ideas as to the nature of the Sun. Philosophers, like the common people, saw in it a globe of fire: the wisest averred that it was "pure fire," "the elementary fire," "the principle of light and fire." But, as there were no means of examining the face of the star, and its real distance from the earth was unknown or imperfectly understood, a prudent reserve was maintained on the subject. The invention of the telescope at once put astronomers in possession of the true celestial domain: it enabled them to fathom the depths of space, and to study the apparent configuration of the stars, including the Sun himself. A few hours observation with an astronomer's glass taught them more as to the nature of the Sun than the two thousand years of dreams, more or less philosophical, which had preceded the discovery of the telescope.

With a glass magnifying the diameter of the Sun not more perhaps than twenty-six times, Galileo, repeating the observations of Fabricius, discovered the spots on the Sun. Although Galileo did not use the black glasses that are now interposed so advantageously before the object-glasses of telescopes, in order to examine without inconvenience the surface of the Sun; and although he limited his observations to times when the sun was on the horizon, rising or setting, or veiled by light clouds, — he studied the spots thoroughly and gave a faithful description of them.

These discoveries by the way greatly astonished the wise men of that day, slaves as they were to the authority of Aristotle. The incorruptibility of the Sun was a sacred principle in the schools, according to the teaching of Aristotle; and these unlucky spots seriously disturbed the philosophers. The Peripatetics strove emulously to prove to the astronomer of Florence that the purity of the Sun was an unassailable principle, and that the spots he had seen were nowhere but in his eyes, or in the glasses of his telescope.

But Galileo had seen clearly, and each of these men could soon convince himself of the reality of the phenomenon that the Florentine had announced.

Not only, in fact, are there spots on the Sun, but they supply the only means we have of learning the astronomical peculiarities and physical properties of that luminary.

The examination of these spots led to the discovery that the Sun turned on his axis, like the planets, making the revolution in twenty-five days. The Sun's days must be therefore twenty-five times longer than ours. It is important to understand the true meaning of the word "day." With us a day is the periodical return of the Earth to the same point after a complete revolution on its axis, with an alternation of light and heat. It is otherwise with the Sun, which, luminous in himselt and in all his parts, can never know night.

As we have said, it was by the examination of the spots on his surface that the rotation of the Sun on his axis in twenty-five days was ascertained. Indeed, if we patiently watch the movements ofone of these spots, or of a group of spots, we find that it proceeds slowly from one side of the disk to the other; leaving the eastern side for instance, it arrives, moving with uniform speed, at the western side, and occupies fourteen days in the journey. If we wait fourteen days more, which are occupied in traversing the opposite, and then invisible face of the solar disk, we see the same spot reappear on the eastern side. The spot, therefore, has occupied twenty-eight days in reappearing. This period of twenty-eight days does not represent the exact duration of the Sun's rotation. We must not forget, indeed, that the Earth does not stand still during this prolonged observation: it has moved around the Sun, and in the same direction as the spots. This advance, which makes us to see the same spot longer than we should have seen it if the Earth remained motionless, amounts to three days, which, subtracted from the aforesaid twenty-eight days, leave twenty-five days as the real time of the Sun's rotation on his axis.

In the Sun, seasons are known no more than days. Time seems to have no existence for the dwellers in that glorious home. The changes and succession of things which make up life for us are unknown to their sublime essence. Duration has no measure in that happy world.

Dwellers in the Sun must see the planets rolling about them, making their revolutions in the same direction and with unequal speed. The phases of the planets and their satellites, of Mars and Venus, or of the moon, that we see from the Earth, they know not: they see of these globes only the hemisphere which is lighted by their luminous land. They see, in immense proportions, Mercury and Venus, and in smaller magnitudes the Earth and Mars. As to the distant planets, Jupiter, Saturn, and Uranus, these must seem very small to them. Neptune must entirely elude their vision. Of the comets they have a protracted view, and see their flaming masses coursing toward themselves and ever growing larger. They see also some comets that have sunk in space, and others that have fallen even upon the surface of the Sun, to be lost and absorbed in his substance.

Thus the spots of the Sun have revealed to us an important specialty of his astronomical duties, — his revolution on his axis. They have also helped us to the only clear ideas that we have as to the physical constitution of the Sun.

We must refer to the illustration to explain what these spots are. Figures 15 and 16 show their general aspect.

In the centre is a dark region utterly unresponsive to our inquiries. Next to it, going from the centre to the sides, is a space marked by half tints, whose gradations melt gently into the rest of the luminous mass. The first region is called the *umbra;* the second, the *penumbra.* It is important to have a clear understanding as to these words. The part known as the *umbra* is dark only relatively to those parts which are brilliantly lighted. This shade (*umbra*) is yet very bright, for its brilliancy has been found to be two thousand times greater than that of the full moon. There is no question here, then, except as to relations of comparison.

These spots are often of considerable size. Some are even 30,000 leagues in extent: they would swallow up the Earth, which is only one-tenth as large. They are not constant: they maintain the same condition sometimes whole months, or even years; but most of them grow or diminish rapidly, and disappear in a few

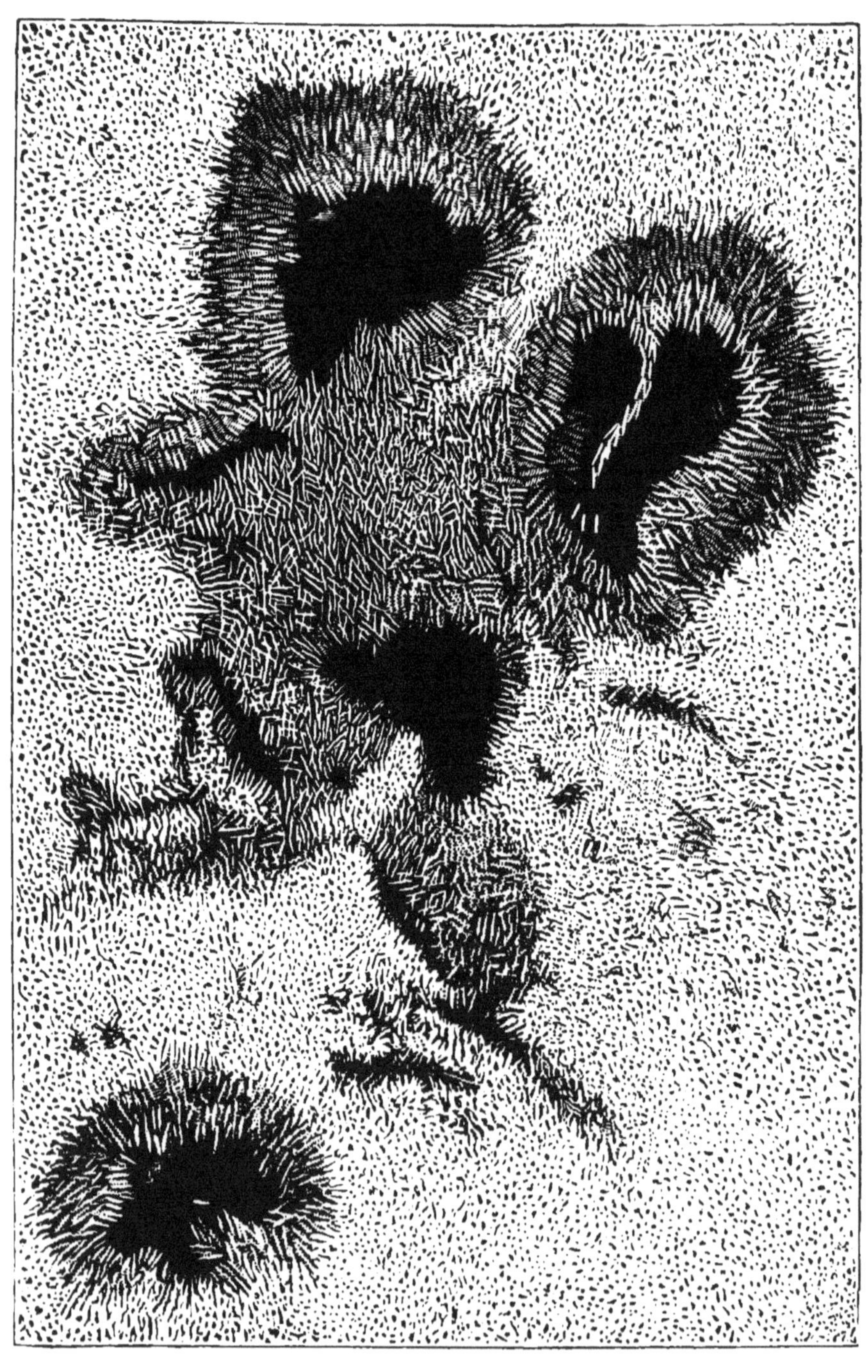

Fig. 15.—Groups of solar spots, seen, in 1864, by Nasmyth.

months. They change their form and extent continually, enlarging or contracting. It is evident that a vio-

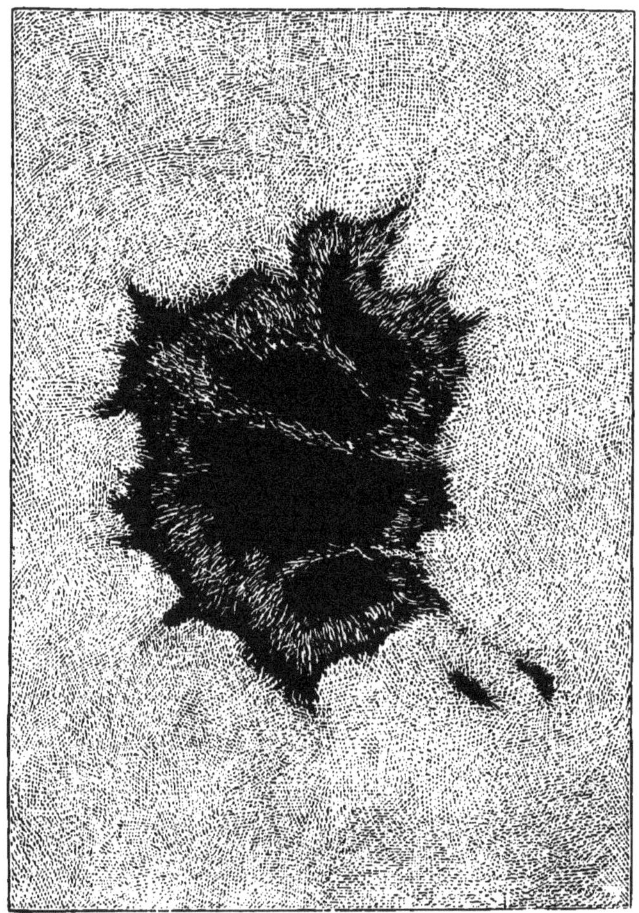

Fig. 16.—Another solar spot seen by Nasmyth.

lent internal motion agitates them, and that they are theatres of tumultuous disturbance. We see what seem to be whirlwinds rush across the regions occupied by these spots and sweep them away, as if in the waves of a furious sea, or in the flames of a conflagration. Gigantic bridges of matter, apparently in flames, have been seen, thrown from one side to the other of two adjacent spots, uniting them by a flashing trail, which soon spread over neighboring spots: by and by the whole structure was overwhelmed in fresh whirlwinds. In a word, there were indications of tremendous agitations, of Titanic throes. These whirlwinds, tempests, and flames are far more prodigious than those of our atmosphere; for the atmosphere of the Sun is many thousand times loftier than ours, and covers thirteen hundred thousand times as much surface.

We have just said that the Sun had an atmosphere. This conclusion, indeed, was reached by careful observation of that luminary.

After the first observations were made of the Sun, a theory of its constitution was formulated, which has come down to our day uncontradicted. In the eighteenth century, the astronomers Wilson and William Herschel developed the theory which, in this generation, has been popularized by the writings of Humboldt and Arago.

According to this theory, the Sun must be composed of a dark nucleus, and a burning atmosphere, which must be the sole source of the light peculiar to this star. Arago and Humboldt called the incandescent atmosphere of the Sun the photosphere. The light, then, would come to us not from the nucleus, but from the photosphere. The spots are explained, on this theory, by supposing them to be openings accidentally made in the Sun's atmosphere, by gases belched from volcanic mouths, or by some other cause. Through these openings must be visible the dark nucleus of the Sun. The *penumbra* of the spots must be formed by the lower part of the Sun's atmosphere, which of itself must be neither warm nor luminous. This lower part of the atmosphere, reflecting the light transmitted by the upper part, or photosphere, must be feebly heated, and only half-lighted.

This theory of the constitution of the Sun and the solar spots for a long time seemed satisfactory. In the same way — that is, by partial eruptions of gas proceeding from volcanic craters — was explained the space of dotted black seen on the surface of the solar disk, and which is faithfully represented in the two figures that we have just seen.

The most brilliant parts that spangle the surface of the Sun, and puncture it here and there with points of intense brilliancy, are called *faculæ*. These are caused, it is said, by certain local accidents, that in certain regions of the solar atmosphere induce evolution of heat and light.

Thus, according to this theory, the Sun must be a body, solid, opaque, dark, like the planets, and enveloped in a first atmospheric *stratum*, which preserves the dark nucleus from too great heat. Above there must be a second atmosphere, the photosphere, which alone must be luminous, and alone enjoy the prerogative of emitting light and heat. A dark nucleus, a dark atmosphere, a photosphere, — these are the constituent elements of the Sun according to Wilson, William Herschel, Humboldt, and Arago.

Even if we adopt this theory, it is not impossible to believe that the Sun may be inhabited by beings not differing much from man, and furnished with an organization analogous to that of dwellers on the Earth and the planets. Shielded, by the interposition of an atmosphere cold and of small conductivity, from the radiation of the photosphere, the body of the Sun is cold, and it is believed that creatures organized not far otherwise than ourselves could live there. The heat of the flaming photosphere may pass through the density of the lower atmosphere only in the degree necessary for sustaining life. The light thus sifted is brilliant, but not dazzling; and it permits the existence of beings organized like those who inhabit the Earth and the other planets.

So Arago did not hesitate to reach this conclusion: "If you ask me this question, Is the Sun inhabited? I should answer that I know nothing about it. But if you ask me if the Sun can be inhabited by beings organized like those who people our globe, I should not hesitate to answer, Yes."

Arago would hesitate to-day; for Science has made large advances in investigating the physical constitution of the Sun. The new method devised by MM. Kirchhoff and Bunsen, and called spectrum analysis, applied to the solar rays, has given birth to ideas wholly new as to the nature of the Sun. It has led us back to the opinions of the philosophers of the Middle Ages, who saw in the Sun a globe of fire, — a kind of gigantic torch.

It would be impossible to go into details of the optical experiments, which have enabled us to make a close analysis of the solar rays, and to deduce from their properties a new theory as to the constitution of the Sun. We will simply state this theory according to the experiments of M. Kirchhoff.

In the opinion of this German philosopher, the Sun is not, as has been declared up to the present day, a body dark, cold, and solid, enveloped in a burning atmosphere: it is a globe, a sphere, probably liquid, burning throughout its entire mass, and in all its parts. This incandescent globe is surrounded by a very heavy atmosphere, formed simply of vapors which proceed from the globe, and which themselves burn in consequence of the high temperature of all these masses of fire.

How, on this theory, can the spots on the Sun be explained? M. Kirchhoff argues that by certain unknown causes a cooling is wrought in the atmosphere of vapors which surrounds this body of the Sun. Thence there must happen condensation of vapor like that of the watery vapors which, on the Earth, produce clouds and rain. These agglomerations of condensed vapors will form in the Sun's atmosphere a kind of clouds; and these clouds, intercepting from us the light of the solar disk, will produce the effect of a spot on the disk. The cloud once formed occasions the cooling of parts of the neighboring vapors, and, inducing a partial condensation round about, causes the appearance of *penumbra*, which circumscribes to our sight the shadow of the spots.

Thus, according to M. Kirchhoff, the solar spots must be clouds suspended in the atmosphere of the Sun. Galileo had previously advanced a similar hypothesis.

Another explanation has been given of the spots, without the sacrifice of M. Kirchhoff's theory. A German philosopher believes the spots to be not clouds in the Sun's atmosphere, but partial solidifications of the liquid matter which constitutes the body of the Sun, — a kind of slag like that seen in crucibles holding matter in fusion, and which comes from the yet unmelted parts of the metal, or parts just beginning to solidify. The *penumbra* of the spots would be explained by the film half-melted, and therefore semi-transparent, which always surrounds with a half-fluid *stratum* the edges of metallic scoria.

M. Faye, a French astronomer, has propounded a theory which changes slightly M. Kirchhoff's. He believes that the nucleus of the Sun is neither solid nor liquid, but wholly gaseous. The solar spots he, like M. Kirchhoff, thinks are openings accidentally made in the Sun's atmosphere by the condensation of vapors on certain of its parts. The spots are caused, M. Faye believes, by currents of vertical vapors ascending and descending: where the ascending currents predominate by their intensity, the light of the Sun's atmosphere is intercepted.

"Through this kind of vista," says M. Faye, "it is

not the solid nucleus of the Sun, cold and black, that we discern, but the gaseous mass, ambient and internal, whose emissive power, in a temperature of the intensest incandescence, is so feeble, compared with that of the luminous clouds whose particles are non-gaseous, that the difference of emissive power is sufficient to explain the striking contrast offered by the two tints observed by our dim glasses."

On the whole, the new theory, sprung from optical experiments made by German philosophers, seems to explain all the observed facts. It is, moreover, generally accepted at the present day. Some differences of opinion there are on points of detail; but astronomers are almost agreed to-day in thinking, with M. Kirchhoff, that the Sun is a body incandescent in all its parts, like a globe in state of fusion, surrounded by a blazing atmosphere, or else, as M. Faye declares, a mere agglomeration of incandescent gas.

CHAPTER XIII.

The Inhabitants of the Sun are purely Spiritual Beings. The Solar Rays are Emanations of Spiritual Beings who dwell in the Sun. The Continuity of Solar Radiation, inexplicable by Philosophers, explained by the Emanations of the Souls of the Sun-people. Sun-worship and Fire-worship among Various Nations, Ancient and Modern.

WE have concluded, from the discussion of physical astronomy in the last chapter, that the Sun is, as MM. Kirchhoff and Faye believe it to be, a mass of burning gas. But, it will be said, if the Sun is a gaseous incandescent mass, or a globe of matter in a state of fusion, surrounded by an atmosphere of burning gas, where do you place its inhabitants, and with what form do you invest them?

We have said that at each promotion in the hierarchy of beings who dwell in the planetary ether, succeeding human individuals, improvements increase, the senses are multiplied, and the intellectual power is greatly extended. As the happy being, once human, rises by successive deaths and resurrections in the scale of inter-planetary existence, he sees diminish in himself the proportion of material substances, which, together with the spiritual principle, make up his glorious individuality. To complete the statement of our system, we must add that, in our opinion, this superior being, when sufficiently improved and exalted by his several incarnations and his many pauses in his vast journey through the heavens, at last arrives at the state of pure spirit.

Reaching the Sun, he is divested of all material substance, all carnal alloy. He is a flame, a breath : he is all intelligence, sentiment, and thought; no impurity mingles with his perfect essence. He is an absolute soul, a soul without a body. The gaseous blazing mass that constitutes the Sun is therefore set apart for these quintessential beings. A throne of fire must be the throne of souls.

We might go further, and argue that the Sun is not only the home and receptacle of souls who have completed the cycle of their wanderings in the world, but is also nothing else than the very assemblage of those souls come from different planets after passing through the intermediate states that we have described. The Sun must be, then, an aggregation of souls.

Since the Sun is the first cause of life on our globe; since it is, as we have shown, the origin of life, of feeling, and of thought; since it is the determining cause of all organized life on the earth, — why may we not declare that the rays transmitted by the Sun to the Earth and the other planets are nothing more or less than the emanations of these souls? that these are the emissions of pure spirits living in the radiant star, that come to us, and to dwellers in the other planets, under the visible form of rays?

If this hypothesis be accepted, what magnificent, what sublime relations may we not catch a glimpse of between the Sun and the globes that roll around him! Between the Sun and the planets there would be a continual exchange, a never-broken circle, an unending "come and go" of beamy emissions, which would engender and nourish in the solar world motion and activity, thought and feeling, and keep burning everywhere the torch of life. See the emanations of souls that dwell in the Sun descending upon the Earth in the shape of solar rays! Light gives life to plants, and produces vegetable life, to which sensibility belongs. Plants, having received from the Sun the germ of sensibility, transmit it to animals, always with the help of the Sun's heat. See the soul-germs enfolded in animals develop, improve little by little, from one animal to another, and at last become incarnated in a human body. See, a little later, the superhuman succeed the man, launch himself into the vast plains of ether, and begin the long series of transmigrations that will gradually lead him to the highest round of the ladder of spiritual growth, where all material substance has been eliminated, and where the time has come for the soul thus exalted, and with essence purified to the utmost, to enter the supreme home of bliss and intellectual and moral power; that is, the Sun.

Such would be the endless circle, the unbroken chain, that would bind together all the beings of Nature, and extend from the visible to the invisible world.

To those who oppose too rigidly the theory that we have just hazarded, we will put one question that will surely puzzle them, because science has never been able to solve it. We would ask them how the heat of the Sun, and the light which is its consequence, are supported? It is clear that the immense quantity of heat and light that the Sun pours into space come from a source that cannot be exhausted, which must be replenished; otherwise, the Sun would be extinguished. As there is no effect without a cause, the Sun must derive from some source the immeasurable amount of forces that he lavishes upon us in his burning rays.

Pouillet has calculated that the Sun, if there were no source of supply for the losses that he sustains, would cool at the rate of one degree in a century. But this estimate is below the truth. Pouillet supposed that the specific heat of the Sun is the most powerful that can be conceived of. This heat is, it is true, unknown; but instead of supposing it to be of maximum power, which proves nothing, we can make it, by hypothesis, equal to that of water, which is well known. Now, by conceding to the Sun the specific heat of water, we rectify M. Pouillet's calculation, and reach the conclusion that the Sun would be extinguished at the end of ten thousand years, if there were no means of repairing his losses. According to Tyndall, whose experiments inspire us with still greater confidence than do those of M. Pouillet, and which are, besides, more recent, "If the Sun were a block of coal, and sufficient oxygen should be supplied to make it combustible at the degree of temperature peculiar to that body, it would be entirely consumed in five thousand years."

Now the Sun has existed millions of years; for the transition rocks of our globe, in which the first living things appear, go back that length of time. Yet its heat has not perceptibly diminished from the remotest ages. The proof that it has not diminished is found in the fact that climates to-day are such as they were in the tertiary or quaternary epoch. In the tertiary and quaternary soils are discovered the same plants and animals that exist in our day. And, to speak of a period less remote, the productions of the soil have not changed at all in the two or three thousand years of which we possess traditional or historical records.

So the Sun has lost none of his heat in millions of years. Where has he got this heat? Where does he get it to-day? By what means is the unchangeable hearth of this powerful star supplied?

Neither physics nor astronomy have been able to give a satisfactory answer to this question. If, for instance, we open treatises on astronomy, and turn to the chapter on "Conservation of the Solar Heat," we find nothing but hypotheses, not one of which is acceptable.

It has been said, first, that, since the Sun turns on his axis in twenty-five days, this movement must occasion a friction of his surface with the medium in which it moves; that is, with ether. But, if this were so, friction must engender the same heat at the surface of the planets, whose rotatory motion, and especially whose courses in their orbits, are much more rapid than the revolutions of the Sun on his axis. Moreover, if we calculate the elevation of temperature which would result from the friction of the Sun against ether, we find that this heat would hardly suffice to maintain the radiation of the solar star for a single century. We can, therefore, make no account of this hypothesis.

Another theory, more logical, has been supported by Mayer, Wollaston, and Thomson: it accounts for the conservation of the solar heat by the fall of meteors upon the surface of the Sun.

A multitude of corpuscles gravitate around the Sun, and approach near enough to be attracted to and fall upon his surface. These are asteroids, which revolve in great swarms around the Sun. A shower of corpuscles, meteorites, may happen on his surface. Their fall would occasion a heavy evolution of caloric by the transformation of their prodigious rapidity into heat; and this caloric, say the authors of this theory, would be sufficient to maintain the entire solar radiation. Let us listen to Mr. Tyndall on this point: —

"It is easy to calculate" [says the English philosopher] "the maximum and minimum of the speed communicated by the attraction of the Sun to an asteroid moving around it: the maximum is reached when the body approaches in a right line with the Sun, coming from an infinite distance, since then the entire force of attraction is brought to bear upon it, without any loss; the minimum is the speed which would be simply sufficient to make revolve around the Sun a body very near its surface. The ultimate velocity of the first body, at the moment when it strikes the Sun, would be 627 kilometres [nearly 686,000 yards] per second; that of the second about 444 kilometres [about 460,000 yards]. The asteroid striking the Sun with the first stated velocity would evolve more than 9,000 times the amount of heat engendered by the combustion of an equally large mass of coal. It is therefore by no means necessary that the substances that fall upon the Sun should be combustible: their combustibility would not add perceptibly to the terrible heat produced by their collision or mechanical shock.

"We have, then, here a mode of generating heat sufficient to return to the Sun its force as rapidly as he loses it, and to maintain on his surface a temperature which exceeds that attainable by any earthly combinations. The peculiar qualities of solar rays, and their incomparable penetrating power, justify us in concluding that the temperature of their source must be prodigiously high: now we found in the fall of the asteroids the means of producing this excessive temperature."

But the fall of asteroids on the surface of the Sun would have had the effect of increasing the mass of that star, and it does not appear that its size has increased since observations were first made of it. These foreign bodies, increasing its mass, would have produced on the orbits of all the stars an acceleration of motion, which, feeble as it might be, would have become perceptible. Now in the more than two thousand years during which observations of the heavens have been made, a perfect regularity has been noticed in the course of the stars of our solar world.

There is another objection to be urged against this hypothesis: it is that it supposes the Sun to be a solid and resisting medium. It is not such a medium, according to the new theory of the constitution of the Sun, which holds that star to be formed of vapors and gas, or, at most, of a liquid sphere. Further proof that this resisting medium does not exist is found in the fact that several comets, among others those of 1680 and 1843, have passed so near the Sun, at their parhelia, that their motions would have been seriously disturbed by a medium even slightly dense. Now the movements of these comets were in no degree affected by this cause: they were seen to appear at the moment indicated by the regular curve of their orbits.

This latter consideration — that is, the absence of a resisting medium in the Sun — appeared so important to one of the authors of this theory, Mr. Thomson, that he abandoned his position as being incompatible with the facts.

A later hypothesis has been proposed to explain the conservation of the solar heat. The substances that form the Sun were not always united in the aggregation in which we now find them. His molecules were at first relatively very distant from each other, and constituted a chaotic or confused mass. Under the influence of attraction, they gradually came together, agglomerated in a nucleus, which has become the centre of attraction of the whole mass. All this is equivalent to saying that the Sun was originally in a nebulous state, and finally entered the condition of continuous and adherent matter.

"The molecules of the solar nebulosity," says Balfour Stewart, "dashing against each other, from the collision heat was produced; as, when a stone is hurled vehemently from a lofty height, the heat is the last form in which the potential force of the stone manifests itself."

There is, on this point, a system of opinions which is generally accepted as explaining the origin of the planets. In thus coming together to form a continuous whole, the elements of the Sun would have changed their physical condition; and from this change would have resulted an enormous evolution of heat, which would be sufficient to account for the origin of the solar source of heat. We know, indeed, that the condensation of matter is always attended by the evolution of heat; and it has been estimated that the diminution of only a thousandth part of the real volume of the Sun would be sufficient to maintain the solar heat for twenty thousand years.

M. Helmholz, author of this ingenious theory, has further calculated that "mechanical force equivalent to the mutual gravitation of the particles of a nebulous mass would be, at the beginning, 454 times greater than the quantity of mechanical force actually disposable in our system. Therefore $\frac{453}{454}$ of the force derived from the tendency to gravitation would be already expended in heat."

The author adds that the remaining $\frac{1}{453}$ of the original heat would be enough to raise 28,000,000° Centigrade the temperature of a mass of water equal to the united masses of the Sun and the planets; a quantity of heat 3,500 times greater than what the combustion of the entire solar system would generate, if the latter were a mass of coal.

These calculations are very interesting, no doubt; but they have the disadvantage of resting on the idea of the Sun's nebulosity, — an hypothesis which would need to be more closely examined before serving as a basis for so important a deduction. Moreover, if the Sun had been heated by a physical cause which has ceased to operate, its heat would have diminished considerably, it may fairly be supposed, since the Sun began its existence. Now, we repeat, the Sun seems never to have cooled. The theory of nebulosity, therefore, has no better foundation than have those which preceded it.

Thus neither astronomy nor physics offer us any satisfictory explanation of the constant maintenance of solar radiation. Common sense tells us that this ever-burning fire must have some aliment. But Science is yet powerless to discern it.

Where Science puts nothing, we put something. In our opinion the solar radiation is sustained by the continual influx of souls into the Sun. These ardent and pure spirits come to take the place of the emanations constantly transmitted by the Sun through space upon the globes that surround him. Thus is completed the unbroken circle of which we recently spoke, and which binds together by continuous links of a common chain all the beings in Nature, and connects the visible with the invisible worlds. This theory as to the conservation of the solar force we can advance with some confidence, since Science has no exact information to give us on the point in question, and philosophy only fills up a gulf between astronomy and physics.

To recapitulate: the Sun, the centre of the planetary assemblage, perennial source of light and heat, which lavishes upon the Earth and other globes motion, feeling, and life, — the Sun is, we believe, the ultimate home of purified souls, perfected, come to the most exquisite stage of subtility. There the souls are shorn of all material alloy: they are pure spirits, living in the midst of a blazing atmosphere and the burning masses that compose the Sun. That star, whose volume far surpasses the united size of all the stars that make up our worlds, is vast enough to give them an asylum.

From this throne of fire, these souls, all intelligence and activity, look down upon the marvellous procession of all the planetary globes that belong to the solar world. In the centre of the world, comprehending all the secrets of Nature and the mysteries of the Universe, they enjoy perfect happiness, absolute wisdom, and knowledge that has no bounds.

Charles Bonnet, the Genevan philosopher, first brought to light the general ideas, of the order of those which we have just developed, as to the philosophy of the Universe. In his "Philosophic Palingenesis," published in 1770, he put forth the doctrine of the human soul's plurality of lives beyond the Earth. In a chapter annexed to that work, and entitled "Faint Conjectures as to the Good to come," he drew a picture of the absolute happiness that we shall taste in that home, and set forth in clear light the transcendent knowledge that we shall possess, and that will unveil to us all the secrets of the physical and moral worlds. Let us quote from these eloquent pages: —

"If the Supreme Intelligence has varied all his works here below; if he has created no two things alike; if a harmonious progression rules among all terrestrial beings; if the same chain embraces us all, — how probable is it that this marvellous chain extends to all the planetary worlds, uniting each to each, and that they are but constituent and infinitesimal parts of the same series!

"We see now only a few links of this grand chain; we are not even sure that we see them in their natural order; we follow that admirable progression but very imperfectly, and with innumerable deviations; we encounter frequent interruptions; but we know well that these gulfs belong far less to the chain itself than to our own powers of comprehension.

"When it shall be permitted us to contemplate that chain, as I suppose those intelligences do for whom our world was mainly created; when we can, like them, follow its prolongation into other worlds, — then, and only then, shall we recognize their reciprocal dependences, their hidden relations, and the proximate reason of each link, and shall rise by a ladder of relative improvements even to the most transcendent and glorious truths.

"With what emotions will not our souls be inundated, when, having fathomed to the bottom the economy of one world, we fly towards another, and compare the economies of the two! How perfect then will be our cosmology! How admirable will be the generalization and fecundity of our principles, the concatenation, the multitude, and the justness of our deductions! What light will flash upon so many different objects in other departments of investigation, — on our physics, on our geometry, on our astronomy, on the natural sciences; and, more than all, on that science which concerns itself with the Being of beings!

"All truths are connected, and the most distant are held to each other by hidden knots: the part of understanding is to discern those knots. Newton, no doubt, gloried in having learned how to disentangle the mysterious connection between the fall of a stone and the movement of a planet: some day, transformed into a celestial intelligence, he will smile at such boy's play, and his profound geometry will be to him only the first elements of an infinite other.

"But the reason of man pierces even beyond all the planetary world; it rises even to heaven where God lives; it contemplates the august throne of the Ancient of Days; it sees all the spheres roll at His feet, and obey the mighty impulse of His hand; it hears the acclamations of all the spirits, and, joining its adoration and its praises to the majestic chants of these hierarchies, it cries out, oppressed with a sense of its own nothingness, 'Holy, holy, holy, is He who is eternal and the only God! Glory to God in the heavenly places! Good-will to man! Oh, how deep is the wealth of Divine Goodness, which stops not with manifesting itself to man on the Earth by facts so many, so diverse, and so touching:

it is willing even to introduce him some day into the celestial home, and to let him drink at the river of delights. There are many mansions in our Father's house; if there were not, his messenger would have told us; He has gone thither to prepare a place for us; He will return, and take us with Him, that we may be where He will be; . . where He will be, . . not in the parvise, not in the sanctuary of the creation, but in the Holy of Holies, . . where He will be; where will be the King of angels and men, the Mediator of the new alliance, the Chief and Perfecter of the faith, who has opened for us the new way that leads to life, who has permitted us to enter in the holiest place, who has brought us to the city of the living God, the heavenly Jerusalem, to the innumerable multitude of angels, to God Himself who is the judge of us all.' . . .

"In these eternal mansions in the bosom of light, it will be a part of perfection and bliss to read the general and particular history of Providence. There initiated to a certain extent into the profound mysteries of His government, His laws and dispensations, we shall see with wonder the hidden reasons of so many general and special events that now astonish and confound us, and cast us into doubts which philosophy does not always dissipate, but concerning which religion always reassures us. We shall con incessantly the great book of the destinies of the world. We shall pause long over that page which treats of the people of this little planet, so dear to us, the cradle of our infancy, and the first monument of God's paternal kindness to man. We shall discover, not without surprise, the various revolutions that this little globe has undergone before taking on its real form, and we shall follow with our eyes those which it is required to endure in the lapse of time; but what will exhaust our admiration and our gratitude will be the marvels of that grand redemption, which yet comprises so many things beyond our feeble reach, but which have been studied and profoundly meditated by the prophets, and into which the angels desire to look even to the very bottom. A word also, on this page, will delineate our own history, and show us the why and the how of its calamities, its experiences, of the privations which often try the patience of the good here below, purify their souls and exalt their virtues, and unsettle and dismay the weak-minded. For us, arrived at so high a degree of intelligence, the origin of physical and moral evil will offer no perplexity: we shall look at them clearly at their source, and in their remotest effects; and we shall be convinced that all that God did was good.

"On earth we see only effects, and these in a very superficial manner: all causes are hidden from us. There we shall see effects in their causes, consequences in their principles, the history of individuals in that of the species, of the species in that of the globe, of the globe in that of worlds, &c. Now we see things but confusedly, and, as it were, through an obscured glass: then we shall see face to face, and we shall know somewhat as we have been known; in fine, because we shall have an incomparably completer and clearer knowledge of the creation, we shall gain a far profounder comprehension of the perfections of the Creator. And how this knowledge — the sublimest, the vastest, the most desirable of all, or rather the only knowledge — will improve incessantly by the most intimate communion with the Eternal Source of all perfections! I can express no more; I can only stammer; words fail me; I would borrow the tongues of angels. If it were possible for a finite intelligence ever to exhaust the Universe, it would yet spend eternity after eternity in contemplation of new treasures of truth; and, after a thousand million centuries consumed in such meditation, it would only have skimmed this knowledge, of which the loftiest intelligence can acquire nothing but the rudiments."

We cannot close this chapter without remarking that the conclusions of Science with reference to the sovereign *rôle* of the Sun in the general economy of Nature is in perfect accord with the religious ideas of the most ancient nations. Fire-worship has prevailed from time immemorial in Asia, and especially in ancient Persia. It was the Persian banks, we know, that divided the first peoples, the Aryas or Arians, who afterwards occupied and peopled Europe. The worship of fire was the first religion of ancient Asia. M. Burnouf, in his "Inquiries into the Science of Religions," has proved this. We will quote some passages from the work of the accomplished Orientalist: —

"Looking about them, the men of that time (the Arians) perceived that all motion in inanimate things, operating at the surface of the Earth, proceeded from heat, which manifested itself either in the shape of fire that burned, in thunder, or in wind; but thunder is a fire hidden in a cloud, and which rises with it in the air; the fire that burns is, before the manifestation, shut up in vegetable matter which serves it for aliment; finally, wind is produced when the air is agitated by a heat which rarefies it or condenses it in withdrawing. In their turn, vegetables derive their combustibility from the Sun, which makes them grow by the accumulation of his heat; and the air is warmed by the Sun's rays. The same reduce the waters of the Earth to invisible vapors, and then to thunder-bearing clouds. The clouds scatter rain, make rivers, supply the seas that the disturbed winds beat upon. Thus all this mobility that animates Nature around us is the work of heat; and heat comes from the Sun, who is at once the celestial traveller, and the universal motor.

"Life also seemed to them to be closely bound up with the idea of fire. . . . The grand phenomenon of the accumulation of heat in plants — a phenomenon which science has since explained — was very early observed by these ancient men: it is frequently mentioned in the Vedas in expressive language. . . . When they lighted wood on the hearth, they knew that they only 'forced' it to give up the fire received from the Sun. When their attention was directed to animals, the close tie which united heat and life in these creatures struck them forcibly: heat nourished life; they found no living animals in which there was life without heat; they saw, on the contrary, that vital energy was exhibited in proportion to the animal's possession of heat, and that the one waned simultaneously with the other. . . . Life exists and is perpetuated on the Earth only under three conditions: that fire penetrates bodies in three forms, of which one resides in the Sun's rays, a second in igneous aliments, and the third in breathing, which is air renewed by motion. Now these last two come from the Sun, each in its way: its celestial fire is, then, the universal motor and the father of life; that which it first engenders is the fire of the Earth, born of its rays; and its second eternal colaborer is the air set in motion, which is also called the wind or the spirit." *

* Revue des Deux Mondes, April 15, 1868.

Sun-worship prevails at this day among all the negro tribes who inhabit the interior of Africa: it may even be said that it is the only religion of the savage Africans, and this religion has existed among them from the earliest time.

The ancient inhabitants of the New World had no other worship than that of the Sun. This is clearly established by the Indian tribes of which historic records are extant,—as the Aztecs, or ancient people of Mexico; and the Incas, or ancient Peruvians. Manco Capac, who conquered Peru, and gave laws to it, was held to be the son of the Sun.

All these primitive people, whose customs date back to the origin of humanity, when they render devout homage to the Sun,—do they not obey a mysterious intuition, a secret prompting of Nature?

However this may be, it is not a little remarkable that the religious ideas of the most ancient peoples should be in such perfect harmony with the latest and most authoritative conclusions of modern science.

CHAPTER XIV.

What are the Relations that subsist between Ourselves and the Superhumans?

HAVING depicted the transmigration of the souls which, parted from man, attain the sublime residence in solar space, we will return to the superhuman, and try to solve a problem which closely concerns us inhabitants of the Earth. We mean to inquire if the superhuman — that is, the being next higher in rank to man, a renewed man, incarnated in a new body, and dwelling in the plains of ether — can put himself in communication with the inhabitants of our globe, despite the vast distance which separates them.

We have already (in Chapter IX.) endeavored to divine the attributes of the superhuman. In view of the number and comprehensiveness of the faculties which seemed to us to belong to him, we have not hesitated to credit this powerful being with the ability of communicating with our Earth and exerting some influence upon it.

But how can such communication be effected? What agent must we presuppose, through whom beings floating far beyond our atmosphere, amid the ethereal expanse, can produce an effect here below? What transcendent electric telegraphy can the superhuman employ? As to this we are absolutely ignorant; yet the mere fact that there is communication between these beings and our Earth seems to us unquestionable. Let us examine the grounds of our conviction.

First, let us speak of public sentiment. As we have already said, we do not hesitate to invoke popular prejudices and opinions, because they almost always embody some great moral truth. Observations repeated thousands and thousands of times, traditions transmitted from generation to generation, and which have endured unchanged and indestructible on the roll of Time, cannot be fallacious. Only, it must be added, since the people among whom tradition is born and cherished lack enlightenment, they hand down their observations in a crude form. But learn how to divest vulgar opinions of their material husk, and you will find a certain truth within. What are ghosts, the idea of which is so firmly rooted in the imaginations of most civilized people? Pluck away the ridiculous white sheet and the form of humanity with which silly rustic superstition invests ghosts, and you will find under all the idea of communication between the souls of the dead and the living; that is to say, the thought that we are trying to express here under a scientific form.

The same popular notion of ghosts we find magnified and shared by persons enlightened apparently, but in fact quite as ignorant of philosophy as those simple dwellers in the fields, and moreover given over to a mysticism which blinds their intelligence and shuts out reason.

The devotees of a new superstition, which sprang up in Europe and America about 1855, in the train of the moral malady of "table turning," are called "Spiritualists." These good folk imagine themselves able at will or caprice to bring back to earth the souls of the dead, those of distinguished men or of their own relatives and friends. They summon the soul of Socrates or Confucius, as well as those of their deceased parents, and fondly believe that these souls come at their call to converse with them. One person, known as the "medium," acts as intermediary between the summoner and the summoned. The "medium," under the spell of an hallucination habitual to him, and of which he is unconscious, writes on paper the answers made by the invoked soul, or rather writes all that passes through his poor head; really imagining that he is conveying messages transmitted from the other world. The listeners accept as a revelation from beyond the grave what is merely the thought of the ignorant "medium."

There is a true and respectable idea in "Spiritualism," — the possibility of man's putting himself in communication with the souls of the dead; but the gross means employed by the partisans of this mystical doctrine lead every enlightened and reasonable man to repudiate all affiliation with them. We mention "Spiritualism" here only as a vapid and vulgar expression of the popular notion about ghosts. "Spiritualism," no doubt, has higher pretensions; but we can grant it nothing more, while we have any respect for science and reason.

The fact of communication between superhumans and the inhabitants of the Earth being, as we think, proved, we shall now inquire how it can be effected.

It seems to us that mainly in the state of sleep and by the agency of dreams this communication is effected, and for this reason: Sleep, that state so strange and so imperfectly explained, is a condition of our existence in which some of our physiological functions — those which connect us with the external world — are done away with, while the soul maintains a part of its activity. In this state, while the body is smitten with a kind of death, the soul, on the other hand, often continues to act, to feel, and to manifest itself by the phenomenon of dreams. Now in the superhuman the spiritual element, the soul, largely predominates over the material. The superhuman is, so to speak, all intelligence. Man when sleeping and dreaming is brought nearer to the superhuman than when he is awake: there is then a stronger resemblance, a closer natural affinity between them. Consequently communication is more possible between these two beings related by kinship of condition.

There is a phrase in the language which is logically sound, and which is the result of numerous and repeated observations. It is said that "sleep brings counsel." Does not this mean that in the night we receive secret communications and salutary advice from the invisible and beloved beings who watch over and inspire us with their supreme wisdom? One thing is certain, that, when we have a decision to make, an idea to discover, we often go to sleep oppressed with perplexities and doubts, and wake in the morning with the decision or the idea clearly and fully attained. That is what this saying means: "Night brings counsel."

The ancients and the people of the Middle Ages attached great importance to dreams. These were believed to be the work or the warnings of God: hence the pains taken to explain them. "In sleep," says Tertullian, "are revealed to us the honors that await men; in sleep remedies are indicated, thefts detected, treasures discovered."*

---

* Liber de Anima, chap. xlvi.

Visions were invested with great significance by Christians of the Middle Ages. It was in sleep that saints, prophets, and devotees received extraordinary communications.

We are far from arguing that only in sleep and dreams can we feel the presence and influence of superhuman beings. There are few persons who have not felt that influence in their waking hours, without accounting for it. They feel as it were a soft and gentle impression, a kind of vague and mysterious touch, that fills their souls with a courage before unknown, a sudden inspiration, an unhoped-for suggestion.

It should be added that all men are not qualified to receive these mysterious impressions. The superhuman can manifest himself only to those whom he loves, and who hold him in reverential memory, and those whom he desires to guard against the snares and dangers of earthly life. It is a mother or a father, snatched from filial affection, come to speak to the soul of him who remains on earth sorrowing in his bereavement. It is a son plucked in the morning of life from the arms of his parents, come to console them for his loss, to enlighten them with his counsels, to furnish them from his profound wisdom with the means of enduring the trials of this life here below. It is two friends who meet again, despite the barrier of the tomb. It is two lovers who, though separated by death, are yet reunited. It is the adored wife torn from her unhappy husband, come to reveal herself to his heart. Then are born again all the sentiments of mutual affection that once subsisted between them: death, which seemed to have broken the bonds that linked their souls together, only veiled them from indifferent or stranger eyes.

Yet, in order to receive these precious communications, a man must have a pure and noble soul, and must have maintained his reverential tenderness for those whom he has lost. The mother who has been indifferent toward her child in his life, or soon forgotten him after his death, cannot hope for these secret messages from him for whom she cherished an imperfect affection. He who has suffered the image of his dead friend to fade from his heart must renounce these sweet manifestations. Besides, however strong may be his affection, however constant his memory of the lost, the man given up to low instincts, to perverse propensities, cannot flatter himself with the prospect of receiving such communications. Only a creature holy, pure, and noble can hold correspondence with these favored beings.

There is a power in our hearts that no philosophy has been able to explain, no science to analyze : it is called Conscience. Conscience is a holy light burning within us, that nothing can ever smother or obscure or quench, and which enlightens us without the possibility of error in all circumstances of life. Conscience is indeed infallible. In spite of every thing, despite all our own interests, apparent or real, always and everywhere, with the great and the humble, the potent and the feeble, she makes us to distinguish the good from the bad, the right way from the wrong. Conscience, we believe, is simply the influence transmitted by a being who was dear to us, and of whom death has bereft us. It is a relative, a friend, gone from the Earth, who deigns to reveal himself to us, to guide us, to show us the better path, and study our happiness.

There are perverse, cowardly, base, and lying men. It is said of such that "they have no conscience." They have not, in fact, the inner light; they cannot distinguish good from evil; the moral sense is wanting in them. It is because they have never loved any one; and their souls, low and vile, deserve not the visitations of one of these superior beings, who show themselves only to men who resemble them or who loved them. A man without conscience, then, is one who, by the viciousness of his soul, is rendered unworthy of supreme counsels, and of the guardianship of those who have gone before.

But it may be noted that this idea of a supreme and invisible protector of man, who directs his heart and enlightens his reason, has already been formulated by the Christian religion, which has borrowed it from the Holy Scriptures. This is the "guardian angel," — the symbol mysterious, poetic, and charming, the seraphic creature whom God has charged with watching over the Christian, to guard him against every snare, to guide him always in the paths of holiness and virtue. We point out this accordance without having sought for it. We record, in fact, such of our ideas as are logically deducible from each other, and without assuming any thing. And, when we find ourselves led into a doctrine of the Christian religion, we are happy to note this accordance.

We invite those who have read these last pages to interrogate themselves, to combine their recollections, to reflect upon what has happened around them; and we are sure that they will also discover many facts in harmony with what we advance. The phenomenon of influences exerted by the dead upon the minds of living persons who loved them, or who cherish their memory, is one of those truths that every one knows by intuition, so to speak, and which every one recognizes as a truth when he sees it formulated and demonstrated. We cannot substitute ourselves for our readers in calling up facts of this kind that must be known to them: we can only report those which have come within our own knowledge. Here they are, briefly stated:—

One of our friends, Count de B——, an Italian, lost his mother nearly forty years ago. He assured us that he had never failed, a single day, to hold communication with her. He added that, to the constant influence and the secret counsels he received from his dead mother, he owed the favorable course of his life, his works and his career, and the good fortune that always attended his enterprises.

Dr. V——, a professed materialist, who, according to the common phrase, "believed in nothing," nevertheless believed in his mother. Like Count de B——, he had lost her at an early age, and had never ceased to be conscious of her presence. He told us that he was oftener with his dead than he had been with his living mother. This declared apostle of medical materialism had, without knowing it, conversations with a soul that had vanished from earth.

A celebrated journalist, M. R——, lost a son twenty-four years of age, a charming and amiable youth, a writer and a poet. Every day M. R—— held conversation with his dead son. Fifteen minutes' solitary meditation placed him in direct intercourse with the being who had been snatched from his embrace.

M. L——, a lawyer, sustained similar relations constantly with the sped soul of his sister, in whom, according to his statement, all human perfections were blended, and who never failed advantageously to advise her brother in all his troubles, great and small.

Another consideration supports the idea under discussion. It has been remarked that artists, writers, and thinkers, after the loss of one dear to them, have been conscious of an increase of power and inspiration. It might be said that the intellectual faculties of the lost one have reinforced their own powers, and doubled their forces.

I knew a financier who possessed very remarkable business capacity. When a difficulty confronted him, he paused, not troubling himself to seek for its solution. He waited, well knowing that the idea he wanted would come to him, without his knowing it. And either in a few days, or a few hours, the expected idea came to him, indeed, spontaneously. This fortunate and admired man had suffered one of the deepest griefs that the heart can know: he had lost his only son, a youth of sixteen years, in whom were summed up all the qualities of mature age and all the graces of youth. Draw the inference for yourself, reader.

The last-cited example may throw some light on these manifestations of superior beings that we are inquiring about. We have said that a certain time was required — some days, perhaps — for the production of these manifestations. This is because the superhuman who makes them has surely many difficulties to encounter before he can put himself in communication with the inhabitants of our globe. It often happens that there is more than one being on Earth whom he loves and wishes to protect, and he cannot be in two places at once. We believe, indeed, that the obstacles that superhumans meet in seeking to open communication with us, together with the sight of the suffering and misfortune that burden their friends here below, are the causes of the only sorrows that they feel in their life, so wonderfully blissful in all other respects. Perfect happiness cannot be in this world; and Fate can still pour a drop of gall or bitter dregs in the cups of happiness drained by the dwellers in ether, in their celestial home.

Persons who have received communications from the dead have made a remark which should be noted here: that communications sometimes cease very suddenly. A famous comedy actress, now retired from the stage, had unmistakable converse with some one who had perished in a tragical manner. She saw the communications suddenly cease. The soul of the regretted dead warned her of the approaching discontinuance of the connection. The reason it rendered serves to explain why this intercourse is subject to interruptions. The superhuman in correspondence with the dweller on Earth had ascended a grade higher in the celestial hierarchy: he had undergone a new metamorphosis, and now he could no longer hold intercourse with the Earth.

Without multiplying these considerations, we will add that among our French peasantry communication with the dead is of common occurrence. In the country death entails none of those lugubrious ideas that it inspires in residents of cities. The memory of the lost is cherished, loved: they are deemed happy whom a kind Providence has early rescued from the woes, the bitterness, the enfeeblement, and the decay of earthly life. They are called upon, and made intimates; and the dead, grateful for this pious recollection, answer to the simple appeals of their petitioners.

All the Orientals have this serene aspiration toward death, which in Europe is exclusively possessed by country people. The Mussulmans love to invoke, to awaken everywhere, this idea of death. The melancholy proverb of the Arabs is familiar: "It is better to sit than to stand; it is better to lie than to sit; it is better to be dead than to be living."

The last chapter ended with a quotation from Charles Bonnet, the first philosopher who advanced the doctrine of the plurality of human existences beyond this

Earth. We will close this chapter with some passages from another natural philosopher, a contemporary of Bonnet, and who has defended the same doctrine with great ability. Dupont de Nemours, in his work "The Philosophy of the Universe," thus expressed himself as to the possible communication between us and superior beings, the invisible dwellers in other worlds, whom he calls angels and genii: —

"Why have we no clear knowledge of those beings, of whom the congruity, the analogy, and the necessity of the Universe throw a reflection, which alone can point them out to us? of those beings, who must surpass us in improvements, in faculties, in power, inasmuch as we surpass animals of the lowest class and plants? of those beings who must have among them a hierarchy as greatly varied, as gently graduated, as that which we admire among other living and intelligent beings whom we excel and hold subordinate? of those beings whose several orders may be our companions on the Earth, just as we are companions of animals which, lacking sight, hearing, smell, feet and hands, know not who we are even when we do them good or evil? of those beings, of whom others, perhaps, journey from globe to globe, or, even more exalted, from one solar system to another, more easily than we go from Brest to Madagascar?

"It is because we have not the organs and senses that our intelligence needs, in order to communicate with them.

"Thus worlds embrace worlds, and thus intelligent beings are classified, all composed of one substance, which God has more or less richly organized and vivified.

"Such is the probability; and, speaking to vigorous spirits which bend not before strong thoughts, I will dare to say that such is the reality.

"Man can calculate that he often has an interest in being useful to other species; and what is more precious, more moral and amiable, he has an interest in serving them for his own satisfaction, without any other motive than the pleasure he finds in the work.

"Well, what we do for our younger brothers, with our very feeble intelligence and very limited goodness, the genii, the angels (permit me to employ common names to designate beings whom I imagine, yet do not know), those beings, so much better than we, must do, and probably do, for us with more benevolence, greater frequency, and vaster comprehensiveness on occasions that move them.

"We know well that there are intelligences; and it matters little that they are, if one pleases, formed of a kind of matter, composed of a mixture, or without mixture. Their quota of intelligence is very bright, very conspicuous, very clearly demonstrated, very evident: it determines instantly the measurable, ponderable, calculable, analyzable properties of inanimate matter.

"To ascertain the influence in the world and upon us of these superhuman intelligences that can be known to us only by induction, reasoning, and comparison of what we are with other animals, even moderately intelligent, efficiently served by us, and who have not the least idea of us, we must push the analogy farther.

"These intelligences are above us and beyond the reach of our senses, only because they are endowed with a greater number of senses, and with a more fully developed and active life. They are beings better than we are, having more organs and faculties: they must, therefore, in unfolding their disposable powers at will, just as we employ ours, be able to arrange, to work, and manipulate inanimate matter, and to act, as well among themselves as toward intelligent beings who are their inferiors, with greater energy, celerity, light, and wisdom than do we toward the beasts which are subordinate to us. It is, then, conformable to the progress and laws of Nature that superior intelligences can, when they please, render us services at once the most important and the least known.

. . . "These unknown guardians who watch over us, and whom we see not, have not our imperfections, and must prize more highly than we do what is beautiful and good in itself.

"We cannot, therefore, expect to conciliate these intelligences of a higher grade by acts that even man would scorn. We can hope no more to deceive them, like men, by a hypocritical exterior which only renders wickedness more detestable. They can oversee our most secret deeds: they understand our soliloquies, even those that are unspoken. We know not how many ways are theirs of reading what passes in our hearts, whose misery, coarseness, and folly abridge our means of knowing how to touch, to see, to hear, and sometimes to analyze and conjecture.

"The house that a celebrated Roman would have built open to the view of all passers exists, and we live in it. Our neighbors are the chiefs and magistrates of the great republic, invested with the right and power to reward and punish even meditated acts, which to them are not secret. And those who divine most clearly the least variations, the faintest inflections, of our intentions, are the most powerful and wisest.

"Let us then strive to deal, so far as is in our power, with those in comparison with whom we are insignificant, and especially let us understand our own littleness. If it is important that we should admit to our close friendship, to our full confidence and constant society, only the chosen few; if the sweet strife of affection. zeal, good-will, and ability, which is incessantly renewed between them and us, helps to improve us daily, — how great will be our profit, if we give them, so to speak, better and more perfect assistants, not influenced by our interests or our passions or our errors, and in the presence of whom we could not keep from blushing!

"These do not change, they do not desert us, they never vanish: we find them as soon as we are alone. They journey with us, share our exile, our prison, our dungeon: they leap about our necks, thoughtful and tranquil.

"We can interrogate them, and as often as we make the trial one would say that they answer us. Why should they not do so? Our absent friends do this for us, but only those for whom we cherish a warm regard. We can experience something like this in the case of an imaginary person, if he seems to us to embody many heroic and good qualities. How often in perplexing circumstances, in the combat of conflicting passions, have I asked myself: 'What would Charles Grandison do in this case? What would Quesnay think of it? Would Turgot approve? What would Lavoisier advise? How shall I win the approbation of the angels? What course will most nearly conform to the order, the laws, the beneficent purposes of the majestic and wise King of the Universe?' For we can in this way carry even to God the salutary and pious invocation, the homage, the enthusiasm, of a soul eager to do right, and careful to avoid dishonor."

CHAPTER XV.

What is the Animal? The Soul of Animals. The Migration of Souls through the Bodies of Animals.

WE have thus far left animals out of our scheme; although, by reason of their immense number, and their influence on the media that they inhabit, they play a part of the highest importance on the Earth. The time has come for naming the place that we assign to them in our system of Nature.

Have animals souls? Yes: we believe animals have souls; but the soul by no means exercises the same degree of activity in all classes of them. This activity varies in the dog and the crocodile, in the eagle and the grasshopper. The soul exists only in the germ in the inferior animals, zoöphytes and mollusks. This germ develops and expands as animals rise in the series of organic perfection.

Sponge and coral are zoöphytes (animal plants). In these creatures the animal character is obscure and hardly perceptible, though its existence is unquestionable. Voluntary motion, which is the distinctive characteristic formerly adduced in animals, is wanting in these: they are immobile, like plants. Yet their nutrition is the same as that of animals: they should therefore be classed with animals. We cannot, however, accord them a complete soul, but only the germ, the beginning of one. In mollusks, like the shell-fish of land and sea, — the oyster, the snail, the medusa, &c., — motion and the mode of life are governed by the will; and this fact is sufficient, we think, to prove that they have souls, though very imperfect and rudimentary ones. In vertebrate animals, especially in insects, the acts which indicate a reasoning faculty, a deliberation and an action which results from reason, are very many, and repeated every minute. They betray an intelligence already active.

The littleness of the bodies of animals constitutes no argument against the fact of their intelligence. In Nature there is nothing great and nothing little; the enormous whale and the invisible plant-louse are equal before the law: each has received the share of intelligence proportioned to its needs, and the degree of mind in living creatures should not be measured by the scale of magnitude. Every one knows the wonderful intelligence manifested by bees in a hive, and by ants in their encampments and hills. The habits of these two kinds of insects, investigated and demonstrated in our own generation, astonish and almost stupefy us. But bees and ants cannot be exceptions in the class of insects. It is very probable that, in all, intelligence exists in the same degree as in bees and ants; for it does not appear why two kinds of hymenopteral insects should monopolize this advantage, while other kinds of the same order, and other orders of this class, of insects are deprived of it. The fact is, the bee has been thoroughly studied, because it is under our hands as an object of agricultural industry; and therefore man is deeply interested in the acquiring of a knowledge of its habits. For this reason the difficulties which surrounded the study of bee-nature have at last been overcome.

We may add that the observer to whom we owe the disclosure of the habits of bees — the Genevan, Pierre Huber, who published his excellent books at the close of the last century — was blind; and in all his observations was forced to employ the eyes of an illiterate servant (François Burnens), — a fact which proves that this kind of investigation was not positively difficult.

The habits of other kinds of insects, which are still unknown to us, must, according to this, conceal marvels quite as great as those that Huber revealed to us in ants and bees.

We must conclude that insects have souls, since intelligence is one of the faculties of the soul.

We would apply the same reasoning to fishes, reptiles, and birds. In these three classes of animals intelligence gradually perfects itself: the faculty of reason is manifest, and the degree of intelligence seems to advance progressively from the fish to the reptile, from the reptile to the bird.

The mammiferous animals represent a palpable superiority of intelligence over the classes of animals that we have just named.

Yet ought we to copy the degree of intelligence in mammiferous animals from the orders that naturalists have established in this class? Ought we to say that the power of intelligence increases according to the zoölogic scheme of Cuvier; that is, that it rises from cetaceous animals to ferine, from ferine to rodent, from rodent to pachyderm, from pachyderm to ruminant, &c.? Clearly not. It would be absurd to bestow on animals diplomas for intelligence, awarded according to the places they occupy in zoölogical classification. We have no certain means of making such an estimate of detail. We stand on the terms of an accepted philosophic thesis, in declaring, generally, that the intellectual faculties of animals increase, from the mollusk up to the mammifer, following nearly the progressive scale of zoölogical classes; but to go into detail of the order would be to encounter certain contradiction. The soul exists as a germ in zoöphytes: this germ develops and grows in mollusks, and then in vertebrates and fishes. The soul acquires certain faculties, more or less obscure and dim, when it enters the body of a reptile; and these are augmented perceptibly in the body of a bird. The soul is endowed with power still more improved when it reaches the body of a mammifer. Such is the general purport of our system.

Let us go now to the very end of this system. We have declared, in the early pages of this book, that the human soul, at the close of its earthly existence, passes into planetary ether, where it takes up its residence in the body of a new being, morally and intellectually superior to man. If this theory is correct, if this migration of the human soul into the body of a superhuman really takes place, analogy forces us to establish a like relation between animals, and then between animals and man.

We firmly believe, indeed, that there is a transmigration, a transmission of souls, or the germs of souls, through the whole series of animal classes. The germ of the thinking soul which existed in the zoöphyte and the mollusk goes, at the death of these creatures, into the body of a vertebrate. In this first station on its journey, the soul improves and betters itself. The nascent soul gains some rudimentary powers. When from the body of a vertebrate this germ of a soul reaches a fish or a reptile, it undergoes further improvement, and its power grows. And when, escaping from the body of the reptile or fish, it takes on the material form of a bird, it receives new impressions, which beget still other improvements. At last the bird transmits to the mammifer the spiritual element, already magnified and modified. From the mammifer, in which it has re-enforced itself, and seen its faculties multiply, the soul enters the body of a man.

It would be impossible to specify the particular mammifer from which the soul must transfer itself to the body of a man. It would be impossible for us to determine if, before reaching man, the soul has passed through the bodies of many mammifers in succession, each one of more complete organization than the last, — if it came through the body of a cetaceous, then through that of a ferine, then through that of a quadrumane, the last term in the animal series. An attempt to go into detail would be dangerous to a system like ours. To contend, for example, that it is the quadrumane that transmits a soul to us, would be incorrect.

The intelligence of this animal is inferior to that of many others placed higher than it in the zoölogical scale. Apes, which compose only a single family in the very numerous order of quadrumanes, have but mediocre intelligence. Malicious, crafty, and rude, they resemble man only in some facial points, and this resemblance appears in only a very few species. In all other respects the quadrumane is pre-eminently bestial.

Not in the quadrumane, therefore, must we look for the soul that is transmitted to man. But there are animals of high and noble intelligence, who seem to have a strong title to such honor. These animals, however, vary in different parts of the world. In Asia, the wise, noble, and dignified elephant is perhaps the custodian of the spiritual principle that is to pass into man. In Africa, the lion, the rhinoceros, the many ruminant animals that throng the forests, may be the ancestors of human peoples. In America, the horse, the proud dweller on the Pampas, and the dog, in all sections the faithful friend and devoted companion of man, are, perhaps, charged with elaborating the spiritual principle that, transmitted to a child, is going to develop and expand in him, and to become a human soul. A modern writer has called the dog "a candidate for humanity:" he spoke more justly than he knew.

It will be objected that man cannot receive the soul of an animal because he has no recollection of such a genealogy. We shall answer that the power of memory is wanting in an animal, or is in it so fleeting as to be virtually nothing. The child, then, must receive from the animal a soul devoid of memory; and, in fact, the child utterly lacks this faculty. At the instant of his birth he differs not at all from an animal, so far as concerns his soul-faculties. Only at the end of about twelve months does memory gradually appear in him: it is afterwards improved by education. How, then, can the child recollect the life which he led before his birth? Do we remember the time we spent in the maternal bosom?

We will add that the progressive order that we have just marked out for the migration of souls through the bodies of different animals is precisely that which Nature has followed in the first creation of organized life on our globe. We have seen, in the chapter in which geological discoveries are recapitulated (Chapter VII.), that plants, zoöphytes, mollusks, and vertebrates were the first living things to appear on the earth. Next came fishes, next reptiles. After these latter came birds, and later mammiferous animals. Man arrived last on the Earth. Thus our system corresponds with the course of Nature in the creation of plants and animals.

Such is the system that we have devised to explain the *rôle* of animals on our globe, and we doubt not that there is a like filiation of souls from animals to man in the other planets — Jupiter, Mercury, Venus, &c. — as on the Earth. In them, as with us, man must receive the principle of intelligence from an animal.

The basis of this system, it will be observed, is the intelligence that we concede to animals. We put aside in this the generally accepted doctrine which denies intelligence to animals, and substitutes for it a certain obscure faculty called Instinct. But this theory gives the reason of nothing: it puts a word in place of an explanation. It attempts to solve with a mere term of language one of the grandest problems of Nature. The timid and hackneyed philosophy of our day has hitherto agreed to evade great difficulties in this way; but the time seems to have come for fathoming more profoundly the problems of Nature, and refusing to be content with words instead of things.

The ancients did not hesitate to grant intelligence to animals. Aristotle and Plutarch express themselves clearly on this point: they doubt not that beasts reason. In modern times the most illustrious philosophers, Leibnitz, Locke, Montaigne, — the most eminent naturalists, Charles Bonnet, Georges Leroy, Dupont de Nemours, Swammerdam, Réaumur, and others, — conceded to animals the possession of intelligence. Bonnet understood the language of many of them, and Dupont de Nemours has given us a translation of the songs of the nightingale, and the dictionary of the crows' language. It is not easy to understand how, in our day, the contrary doctrine has prevailed; how Descartes and Buffon, who vehemently denied the intelligence of animals, have succeeded in turning the scale in their favor.

Be that as it may, Descartes believed animals to be mere machines, like automata furnished with machinery, and which act only by the working of their mechanical apparatus. It is difficult to be more absurd than is our great philosopher when he reasons about animal machines.* *Equidem bonus dormitat Descartes.* The systematic errors of Buffon on the same subject are well known.

It is the partisans of Descartes and Buffon who have popularized the idea of instinct, set in the stead of intelligence,—the word replacing the thing. But, in good faith, what difference is there between intelligence and instinct? None at all. These two words simply represent different degrees of the same faculty. Instinct is merely intelligence somewhat weakened. Read the writings of naturalists of our day that deal with the question,—Frederic Cuvier, the brother of Georges, and Flourens,† who only annotated the work of Frederic Cuvier, or the profounder treatise of a contemporary naturalist, M. Fée, of Strasbourg, ‡—and you will readily discover that no fundamental distinction can be established between intelligence and instinct, and that the whole secret of our philosophers and naturalists lies in calling "instinct" the intelligence, feebler than ours, which is peculiar to animals.

It is the pride of man that has undertaken to rear between us and the animal a barrier which has no real

---

* In his "Discours sur la Méthode" Descartes treats particularly of this question.

† De l'Instinct et de l'Intelligence des Animaux. By P. Flourens. 4th ed. Paris, 1861.

‡ Études Philosophiques sur l'Instinct et l'Intelligence des Animaux. By M. Fée, Professor of Natural History in the Medical Faculty of Strasbourg.

existence. The intelligence of animals is less developed than man's, because their needs are fewer, their organs less perfect, and the sphere of their activity narrower; but that is all the difference. And sometimes — let us not forget this — the animal surpasses the man in intelligence. Look at the rough and brutal carter by the side of the gentle and docile horse that he loads with blows and furiously abuses, while the faithful servant does his work calmly and completely: look, and say if it is not the master who is the brute, and the animal who is the intelligent being. In point of kindness, that sweet emanation of the soul, animals often surpass man. You know the story of the man who went to a river to drown his dog. His foot slipped, and he fell into the water, and was drowning. But his companion that he had cast to death was there: he sprang to the aid of his imperilled master, and brought alive to the bank the man who meant to be his murderer. The latter, however, more cautious this time, again seized his savior and threw him into the water.

Thus, according to our system, the human soul proceeds from an animal belonging to the superior orders. Having undergone, in its body, partial elaboration and suitable improvement, it goes to be incarnated in the new-born body of a child of man.

We said in a former chapter: "Death is not an end, but a change: we do not die, we undergo a metamorphosis." We must now add: "Birth is not a beginning, it is a sequel. To be born is not to commence: it is to continue an anterior existence."

For the human species then, properly speaking, there is neither birth nor death: there is only a series of existences linked together, and which reach from the visible world through space to connect themselves with worlds shut off from our gaze.

CHAPTER XVI.

What is the Plant? The Plant feels. How Difficult it is to distinguish Plants from Animals. The General Chain of Living Beings.

LINNÆUS said: "The plant lives; the animal lives and feels; man lives, feels, and thinks." This aphorism represents the state of science in the days of Linnæus. But since the year 1788 — that is, since the death of the great botanist of Upsala — natural science has progressed, botany and zoölogy have been enriched by innumerable facts and fundamental discoveries, so that the Linnæan formula does not answer to the present condition of the knowledge of organization. We think the following proposition could be substituted for it: The plant lives and feels; animals and man live, feel, and think.

To concede feeling to plants is to transcend the classical rules of natural history. Therefore we think we ought to set forth carefully the arguments and facts which seem to warrant this proposition.

1. The plant has the sensation of pleasure and of pain. Cold, for instance, affects it painfully. We see it contract or, so to speak, shiver, under a sudden or violent depression of temperature. An abnormal elevation of temperature evidently causes it to suffer; for in many vegetables, when the heat is excessive, we see the leaves droop on the stalks, fold themselves together, and seem withered; when the cool of the evening comes, the leaves straighten and the plant resumes a serene and undisturbed appearance. Drought causes evident suffering to plants. Those who read the touching book of Nature with tender eyes know that the plant, watered after a prolonged drought, shows signs of satisfaction. On the other hand, a bruised plant, a tree from which a large branch has been cut, seem to feel pain. A pathologic liquid oozes from the wound; it is like the blood that flows from the wound of an animal; the plant is sick and it dies, unless proper care is taken of it. So feeling persons who love plants avoid cutting the stalks of flowers: they choose rather to inhale their perfume, and to gaze upon their brilliant hues on the unmutilated plant, without wounding by a painful gash the charming beings that they admire.

The sensitive-plant touched by the finger, or only visited by a current of unwelcome air, folds its petals and contracts itself. The botanist Desfontaines saw one which he was conveying in a carriage fold its leaves while the vehicle was in motion, and expand them when it stopped, — a proof that it was the motion that disturbed it. A drop of liquid acid falling on the leaf of a sensitive-plant produces a similar constriction. All vegetables present an analogous phenomenon. Their tissues curl when they are brought in contact with some irritating substance. Rubbing the tops of lettuce will make the juice gush out.

Vegetable sensibility is of the same title with animal, since electricity kills and crushes plants as well as animals, since narcotic poisons put to sleep or kill plants as well as animals. You can put a plant to sleep by washing it with opium dissolved in water. MM. Goppert and Maccaire have discovered that cyanohydric acid destroys plants as quickly as it does animals.

2. Plants sleep in the night. During the day they develop their vital activity; and when night comes, or when they find themselves in darkness, their leaves take a new attitude, which is a sign of repose: they fold themselves together. Reflecting that leaves are so situated in the day-time that their upper faces look toward the heavens, and the lower earthward, and that the lower face, pierced with holes or spiracles, is the part by which absorption and exhalation are accomplished, while the upper, lacking these openings, is only a kind of screen designed to protect the absorbent face,—we shall understand that this horizontality of the leaves may be a regular position of vital activity, and that the nocturnal folding of these same leaves may indicate a state of rest. It is just so that in the night we give up to absolute relaxation our muscles which have been tense through the day.

The sleep of plants, said to have been discovered by

the daughter of Linnæus,— and which, however this may be, was first described in one of the Botanical Treatises of Upsal, and thoroughly explained by Linnæus, — is not a phenomenon limited to some classes of plants. There are few vegetables that in the night and darkness do not fold their leaves and wear an appearance different from that which they present in the day. The sensitive-plant is the standard by which this phenomenon is most frequently shown in all its intensity; but this little leguminous plant only exaggerates that which occurs, in a feebler degree, in almost every vegetable that has aerial leaves.

Let us repeat here what we said in another work touching this phenomenon: —

> "The sleep of plants vaguely suggests that of animals. Strange though it is, the sleeping plant seems to desire to return to the time of its infancy. It curls itself up almost as if it were in the unblown bud, when it slept the torpid sleep of winter, sheltered under its strong scales, or shut up from the wind in its warm down. One would say that the plant tried every night to regain the position it occupied in its youth, as the sleeping animal cuddles and rolls himself together just as he was in the bosom of his mother." *

Can we deny the sensibility of these creatures which give signs of alternate rest and activity, and which accommodate themselves to different impressions from without? Weariness can be only the consequence of an impression felt.

3. Many physiological functions are fulfilled in plants

---

* Histoire des Plantes, Paris, 1865, p. 111.

as well as in animals; and, when we see their number and variety, we find difficulty in understanding that animals possess sensibility, according to the opinion of all the world, while plants, equally in the opinion of all the world, are destitute of it. An ancient philosopher defined plants as "animals that had taken root." We shall see, in examining the numerous functions that operate within vegetables, whether this philosopher was not a very clear-sighted man.

It would be difficult to ascertain what function belongs to the animal that does not belong to the vegetable in a more or less enfeebled degree.

Respiration, for example, is as much the prerogative of plants as of animals. In the latter it consists of the absorption of the oxygen of the air, and the emission of carbonic acid gas and vapor: in plants it consists in the emission of carbonic acid gas and vapor in the night; and in the day-time, under the influence of the Sun's light, in the emission of oxygen proceeding from the decomposition of carbonic acid gas. This function is evidently of the same nature in both kingdoms.

Exhalation is a function common to animals and vegetables. By the spiracles of leaves, as by the pores of the skin in animals, vapor and several gases, differing according to the vital phenomena going on in the interior of the tissues, are constantly evolved.

Absorption takes place in both kingdoms. Place the lower face of a leaf on water, and see how rapidly the latter will be absorbed. Sprinkle a bouquet with water, and freshness will revisit the withered corollæ.

Absorption is even more active in vegetable than in animal tissues.

The circulation of fluids in the interior of plants is effected by a copious and complicated system of canals and vessels of all kinds and sizes, — absorbing, exhaling, tracheal, channelled. There is nothing more various than the arrangement of canals in the interior of plants. This multiplicity of vessels bespeaks a circulating function as complicated as that of animals.

Vegetables, then, have almost the same physiological functions as animals: only we know them yet very imperfectly. It is very strange that while animal physiology is so advanced, vegetable physiology is almost in its infancy. We are familiar with the process of digestion in man and animals; we know how our blood circulates in a double system of vessels, arterial and venous; and we know the central organ, the heart, in which the two liquids that flow through this double system unite their currents. We see and touch the organs of sensibility and motion; that is, the nerves. More than that, we distinguish the nerves of sensibility from those of motion. We know that the centre of nervous action in man and animals is double, and that its seat is at once in the brain and in the spinal marrow.

In a word, Science has thrown its strongest light on all the functions of the animal organism, while vegetable physiology offers us only obscurities. Notwithstanding the many works of naturalists in the last two centuries, we cannot explain with certainty the life of plants. We cannot confidently say how the sap, the blood of vegetables, circulates in their canals. We do not know surely whether a tree grows from the outside or from the inside. All physiological functions in the vegetable kingdom are obscured to us by the thickest veil; and it is only through the difficult lifting of a corner or two, that we can catch a few glimpses, some clear spots, in the darkness which enshrouds these phenomena.

Yet, inexplicable though they are, physiologic functions exist in plants. Looking at these, we cannot believe that the gift of sensibility has been withheld from their possessors. We cannot understand that they should have, as Linnæus contended, life and nothing more.

It will be objected that plants have no nerves, and that in the absence of any organ of sensibility that faculty cannot be conceded to them. We answer that the imperfect state of vegetable anatomy and physiology forbids us to establish any conclusions as to the presence or absence of nerves in plants. We are sure that these organs exist, but botanists have no means of distinguishing them from other organs.

4. The mode of multiplication and reproduction is so strongly analogous in plants and animals that it seems impossible, in view of this extraordinary resemblance in the most important function of all, to refuse sensibility to plants while we grant it to animals.

Let us look at the different modes of reproduction peculiar to vegetables. Reproduction, or rather the fecundation which precedes it, is effected in the plants called phanerogamic by an apparatus of the same typical form as in the animal kingdom; that is, it is composed of a male organ, the stamen, which enfolds the fecundating dust (pollen), and a female organ, the ovarium, supported by a stem called the pistil. The pollen fecundates the ovule contained in the grains of pollen in the ovarium, as the semen of the male fecundates the ovule in the animal egg. The fruits of each fecundation then develop, with the aid of heat and time. The vegetable egg grows and matures like the animal.

Let us add that the analogy between the two modes of reproduction in the two kingdoms does not end here: resemblance may be detected in the details of the function. A peculiar vital activity, a tumidity of the tissues, attended by a local elevation of temperature, is noticed at the moment of efflorescence,—that is, of fecundation,—in certain plants of the Aroid family. A thermometer placed at such a juncture in the huge floral envelope of the arum (vulgarly called wake-robin, or calf's foot) marks an elevation of one or two degrees above the temperature of the surrounding air,—an extraordinary circumstance in vegetables, which are always colder than the outer air. How can we believe that a plant which is the theatre of such phenomena of excitement has no consciousness of these states? The plant like the animal has its amorous seasons; and will you contend that it is unconscious of them? Will you insist that this plant which becomes heated, in which life is highly excited at the instant of fecundation, feels nothing in its inner nature; that it has no more feeling than the stone that sleeps at its feet? We do not believe this: we cannot conceive of life without sensibility; one seems to us the index of the other.

The analogy between the reproductive functions of plants and animals is nowhere more evident or more curious than in a plant which abounds in the waters of the Rhone, and which has received the name of *Vallisneria spiralis.* This plant is *diœcian*, which means that the male and female organs are in two different stalks of the same plant. Now the female flowers are attached to the soil by long stems, which roll up spirally upon themselves. At the moment of communion the spires of the stem unfold, and the female flowers open on the surface of the water. The male flowers not being supported on an elastic stem, like the female, cannot open at the surface of the water. What do they do? They burst their envelopes, and go floating on the surface around the females. After this the current of the river carries away the male flowers *coupées;* and the female stem, folding itself, redescends to the bottom of the river to mature its fecundated ovules.*

---

* In his poem on Plants, Castel has given a charming as well as accurate description of the loves of the Vallisnerie, and it may please the reader to find it here: —

> The impetuous Rhone, beneath his foaming wave,
> For six whole months conceals from us the plant
> Whose lengthening stalk, in love's sweet season, lifts
> Itself above the waves in the face of day.
> The males, till then immobile in the depths,
> Glide toward their lovers, and, all fancy free,
> Form in the quiet stream a mighty fleet.

We do not abandon the idea of the reproductive function in plants, because it is fertile in conclusions that give support to our position. The plants called phanerogamic are not reproduced simply by impregnation through the apparent sexual organs; that is, the pistils and stamens. They are reproduced, also, by grafting, by cuttings and sprouting. Cryptogamic plants, which have no sexual organs, are reproduced either by spores detached from the individual at a certain epoch of vegetation, — as is seen in brakes, algæ, truffles, and others, — or by fragments of the same individual, which, cast into the earth, have the power of germination and reproduction.

Animals, in all their classes, show us all the modes of reproduction: there is not one of these that cannot be found among them. The animal is not multiplied by eggs, internal or external, alone, or through little living likenesses of himself: he is also multiplied, like the vegetable, by sprouting, cutting, and grafting.

Multiplication by sprouting is witnessed in the freshwater polyp. On the body of this animal little buttons come out, which grow and extend. While the button is enlarging, itself puts forth other buttons or sprouts still smaller, which in turn produce still other. All these sprouts or buttons are so many little polyps which

---

It is a fête, one says, or Hymen leads
Along the waves his happy following.
But Cytherea's rites fulfilled, the stalk
Gently withdraws, with readjusted folds,
And sinks, to ripen the impregnate seed.

draw their nourishment from the principal polyp. Having attained a certain size, they detach themselves from the first individual, and constitute so many new ones.

The coral is multiplied in the same way. From the principal branch proceed secondary branches, which originated in a sprout or bud; and these branches, inserting themselves in the main branch, become new individuals. Therefore the coral, viewed externally, resembles branching shrubbery rather than an animal.

Madrepores, other zoöphytes, so strongly resemble shrubbery that for centuries they have been supposed to be marine plants: like the coral, they are reproduced by sprouting.

Of multiplication by cutting, the fresh-water polyp affords an example. Take one of those animals, and cut him into as many pieces as you please: each of these left to itself will become a polyp. These new individuals, subdivided, would produce as many new polyps. Here is real reproduction by cutting, like the cutting of plants; so that the generation of a fresh-water polyp is not different to that of one of our fruit-trees.

It is not only the entire polyp that thus cut to pieces can yield a new polyp: the mere skin of the animal will furnish one or more new individuals. Is not this a vegetable grafting?

The same generation by grafting appears, under other conditions, in the fresh-water polyp. Reunite the animal, end for end, or approximate different parts of the same polyp, or those of several, and you will see them combine so closely as reciprocally to nourish each other and finally form a single individual. Here again is a veritable vegetable grafting effected in an animal.

5. There are other points of resemblance between plants and animals. If they have not been generally noticed, it is because the authors of standard works on natural history have not called the reader's attention to them. We will try to repair their neglect, and to make apparent these analogies between the two kingdoms of Nature.

First is a common and equally astonishing fecundity. Of plants as well as animals, a single individual can give birth to thousands like itself. Vegetables are even more prolific than the superior animals. Trees produce every year, and in some cases for a century. Mammiferous animals, birds and reptiles, produce far less than trees: their broods are less numerous, and are born only during a certain period of the animal's life. The elm yields annually more than 300,000 seeds, and this productiveness can continue a hundred years. Fishes and insects resemble vegetables in fecundity. A tench lays 10,000 eggs per year; a carp, 20,000. Some other fishes produce even a million eggs annually. Of insects, a mother-bee lays 40,000 to 50,000 eggs. To offset this fecundity among animals, may be cited, among vegetables, that of the corn-poppy, the mustard, the brake, which yield incalculable quantities of seed. It should not be forgotten that vegetables are multiplied in many ways, while each animal has, generally speaking, only one mode of reproduction.

What we wish to prove — and it is evident — is that fecundity is the same and equally prodigious in animals and plants.

Let us cite further, in the way of analogy, the size of species which differs widely in both kingdoms, for both produce at the same time dwarfs and giants. There are some animals of monstrous size, such as the whale, the cachalot, and the elephant; and the gigantic reptiles of the ancient world, — the ichthyosaurus, which was longer than the whale, the megalosaurus and the iguanodon, which were as large as the elephant.

To these colossi of the animal kingdom let us oppose the colossi of the vegetable: the enormous sour gourd which covers with its shade hundreds of square metres; the elm which can grow as long as a whale; the *Eucalyphis globulus*, an Australian tree, which it has been attempted to acclimate in Algeria and the South of France; the *Sequiœa gigantea*, the giant of Californian forests.

If these two kingdoms have their colossi, they also have their dwarfs and creatures infinitely little. There are cryptogamic growths that can be seen only with the microscope, and animalculæ which are also invisible except through that instrument. So, if the animal kingdom can range in the scale of magnitude from the whale to the microscopic acarus, the vegetable kingdom has the same decreasing gamut between the sour gourd and mould.

Moreover the same places are inhabited or sought by animals and plants. Both live on the same soil, as if to aid each other reciprocally. The two kingdoms of Nature interlace their branches on every point of the globe. A multitude of places could be cited in which certain plants and certain animals alike delight. The chamois and the maple love the same mountains; the truffle and the earth-worm inhabit the same subterranean regions; the hare and the birch meet in the same spots; the ape and the palm-tree follow, and the ermine and the ginseng accompany each other; the leech and the conferva are associates; the nenuphar grows in the same sweet water with the moth; the cod and the algæ thrive in the same submarine depths.

All vegetables and animals have a native land, but both by human cultivation can be acclimated under foreign skies. The chestnut and the turkey-cock and the peach-tree have forgotten the lands of their birth.

Among animals as among plants, there are amphibious creatures. The frog and the other batrachians, as well as rushes, live on land and water.

Animals and plants can live parasitically. While the animal kingdom has its parasites, like the louse, the chigo, and the acarus, the vegetable kingdom has its lichens and its mushrooms.

Thus the same fecundity, a like variety in size, similar habitations, which implies identity of organization, possibility of transplantation and acclimation beyond their native homes, and of amphibious existence and parasitic life, — all general conditions which suppose analogous organizations, — this is what we deduce from the parallel between plants and animals. How, then, if we grant sensibility to one of these kingdoms, shall we refuse it to the other?

6. Plants, like animals, have their maladies. We do not refer to maladies caused by parasites, like the vine disease, which is due to the *oïdium Tuckeri;* the potato-rot, occasioned by other little fungi; the diseases of wheat, the rose, the olive, &c., caused by certain cryptogamic parasites, which establish themselves on the plant and change the natural course of its life: we speak of morbid affections, properly so called. The pathologic state and its consequences exist in the plant as well as in the animal. An abnormal and feverish arrest or acceleration of the sap in the vegetable, like the arrest or acceleration of the blood of an animal during a fever; certain excrescences of the bark, like the skin-diseases of animals; the abortion of entire organs, and the vicious development of others; the secretion of pathologic fluids which flow outward,— here is a rapid compendium of the maladies to which trees, shrubs, and herbaceous vegetables are subject. A plant that fades too quickly or too often, owing to intense cold or heat, soon becomes sick, and inevitably dies, like an animal exposed to dangerous extremes. A shrub left in a current of cold air can no more live than could an animal kept in the same place. (This has happened, I may say, by the way, to all the plants I have placed in the vestibule of the ground floor of my house.) In a word, a plant is sick or well according to the conditions to which it is exposed. How shall we maintain that this creature in which such changes

occur is the passive subject of them, that it feels no sensation, neither pain nor satisfaction in passing from sickness to health and *vice versa?*

7. Maladies, or some other cause, produce in plants, as in animals, anomalous shapes and irregularities of structure. As there are monstrosities in the animal kingdom, so are there in the vegetable. That branch of science which deals with monstrosities among animals is called Teratology. Geoffrey St. Hilaire has written some admirable papers on the causes which produce monsters in various classes of animals. But in our time it has been seen that a similar department of science should be organized to explain the monstrosities of the vegetable kingdom, and Moquin-Tandore has published a work on vegetable teratology.

8. Old age and death come to plants as to animals. The plant, having withstood the various maladies which threaten it, cannot escape gradual old age; and death necessarily follows this. In process of time, its vessels become tough; and their calibre, having shrunk, is obliterated, and can no longer give passage to the sap or other fluids which must traverse them. Liquids are not sucked up with the same regularity: they no longer transude through the vegetable tissue with the same precision. Stagnating within the vessels, they become corrupt, and their decomposition is communicated to all the vessels that enclose them. Then the vital functions cease to act, and the plant dies.

The same process takes place in animals. The thickening of the vessels, and the obstruction of their calibre, bring on the state of old age, in which the functions are disturbed and slacken; then comes death, the inevitable end of all in either kingdom of Nature.

Thus, on comparison of animals and plants with special application to the lower grades of animals, it is impossible to establish between them a line of precise demarcation. The marks that ancient naturalists adopted to distinguish plants from animals are now known to be worthless, and the distinction becomes more and more difficult on more intimate acquaintance with these creatures. Voluntary motion was once regarded as the distinctive characteristic *par excellence* between the two kingdoms of Nature; but now it is no longer appealed to. Elementary works on botany have much to say of the *Dionæa* fly-catcher, which seizes insects that walk on its leaves, just as the spider catches flies; and of the *Desmodie oscillante*, the leaves of which have a voluntary motion more palpable than that possessed by many animals.

In addition to these examples, taken from standard works, we would ask what becomes of the argument founded on the immobility of plants, when we see zoöphytes fixed to the soil; and, on the other hand, that some young plants, or their germs, like the germs of algæ, mosses, and brakes, have the power of self-motion?

The spores or reproductive organs of algæ, and the fecundating corpuscles of mosses and brakes, have the fundamental character of animality; that is, they are provided with locomotive organs (vibrating hairs), and make movements apparently voluntary. These strange creatures are seen to go and come in the depths of liquids, to try to penetrate cavities, to withdraw, to return again, and finally to enter with an apparent effort.

Therefore German botanists regard these vegetable germs as belonging to the animal kingdom. Considering the fact that only animals have organs of motion, and that the spores of algæ and the fecundating corpuscles of mosses and brakes are furnished with these organs, in the shape of vibrating hairs, they do not hesitate to declare that, from the beginning of their existence, algæ, mosses, and brakes are real animals, which become plants by fixing themselves in the earth and beginning to germinate.

French botanists have not yet dared to enter this path: they are content with calling *antherozoids* the mobile fecundating corpuscles of algæ, mosses, and brakes, and do not venture to decide as to the animality of these creatures.

This is what M. Pouchet says in his work on the Universe: —

"Mobility is manifested spontaneously with remarkable intensity in the pollinic animalculæ of several plants, which have for this purpose special organs, hairs, by the aid of which they swim in all directions in the liquid that hides them.

"Some, real animalculæ-plants, have the shape of eels, and move by the aid of two long filaments in their heads: this is seen in the common chara. Others strongly resemble the tadpole frogs, and whirl about in the cells of mosses.

"Yet it is such creatures, whose organs of locomotion are so plainly visible, and which the microscopic observer sees hop under his eyes as nimbly as mountebanks in their perilous leaps,— it is such creatures that certain botanists obstinately insist, in pure theory, upon regarding as immobile and insensible. Some *savants*, then, have eyes not to see with, have they not?" *

There are, therefore, germs of plauts and young plants that move; and, on the other hand, nearly all the zoöphytes, the sponge, the coral, the madrepore, the finger-fish, the byssus, and others, to which may be added many mollusks (all the shell mollusks), are fixed to the soil. Here, then, the plant would be taken for the animal, and *vice versa*, if voluntary motion were held to mark the distinction between the two.

On the frontiers of the two kingdoms,— that is, when zoöphytes in the animal, and cryptogamia in the vegetable kingdom, are in question,— there is, so to speak, neither plant nor animal: the two kingdoms seem to be merged.

If, prior to the discovery of the fresh-water polyp, one had been shown to a naturalist, he would have found it difficult to classify him. Seeing him multiplied by buds, by sprouts, by cuttings, by grafting, he would, no doubt, have declared that this organized life was a plant. But if he had been reminded that the same creature subsisted on living prey which he could seize and swallow; that he had, for catching it, long and flexible arms with which he formed a kind of snare; that, finally, he swallowed this prey through a digestive canal, our naturalist would be forced to rank the polyp among animals. He could have been made to remark,

* L'Univers, Paris, 1868, p. 444.

however, that the polyp displays the curious power of being turned like a glove, so that his exterior becomes his interior skin; and that, moreover, having been thus turned, he lives, grows, and reproduces himself just as before the turning. Greatly embarrassed in the presence of a fact so unprecedented, our naturalist would doubtless have sought between plants and animals some intermediate kingdom where he could locate this paradoxical creature that could be referred with certainty to neither plants nor animals.

The fact is, classifications are the work of human science: Nature knows them not. We descend by imperceptible degrees from one kingdom to another: we go from the man to the polyp, and from the polyp to the rose-tree by gradations infinitely fine, and on the bounds of the two kingdoms there is an entire series of creatures that can hardly be ranked in either. How often have naturalists hesitated to accept as animals the coral, the sponge, the finger-fish, the gorgons and sea-anemones, and madrepores. Even in our day microscopic observers, who study the creatures peculiar to animal and vegetable infusions, — such as monads, *vibrions*, the various moulds, &c., — are often at a loss to class them in this or that kingdom, and decide sometimes rather arbitrarily to place them among animals or plants.

From all these considerations, from all the facts that have been set forth, we conclude that we cannot contest the sensibility of animals, since we do not think of denying its possession by zoöphytes, the coral, the sponge, the finger-fish, and the madrepore, &c., which it is often so difficult to distinguish from vegetables.

There is a majestic tree, an oak with great branches, growing on the shore of the sea. Not far off a finger-fish thrown up by the waves sprawls on the sand of the beach. A few feet beneath the surface of the water are seen a sponge, a branch of coral, a madrepore. When the icy north wind blows, or a hurricane upheaves the waves, what is the creature that will show itself sensible to the unchained tempest, — an animal or a plant? The sponge, the coral, and the madrepore will remain as indifferent to the fury of the elements as the rock on which they are incrusted, or as the shingle on which the finger-fish spreads his four marble arms. On the other hand, the mighty oak, which covers with its immense branches a considerable part of the bank, will tremble in the blast, will curl his branches and shut his leaves to shelter himself from the icy norther, and by his very attitude you will understand that an unnatural perturbation reigns in the atmosphere. Will you seriously maintain, in this case, that the vegetable feels nothing, and that the animal does? Will you not rather be led to declare, on the contrary, that the tree is the sensible creature, and that the finger-fish, the sponge, and the madrepore have no feeling at all?

Pause at the brink of a standing pool to look for the fresh-water polyp. You will be troubled to distinguish this zoöphyte in the midst of the rushes and reeds which surround it. At last you will discover a kind of membranous tube, hardly a few centimetres * long. Is this the polyp that you are seeking? Is it not rather the culm of a gramineous plant, or a rush? This living stalk, not distinguishable in appearance from an herbaceous plant, is continually fixed in the same place, like an aquatic vegetable. It performs some obscure movements, which consist merely of opening and closing the membranous tube which constitutes its whole being. Sometimes it lengthens, then contracts itself, stretching its secondary stems, — a kind of membranous arms, fine as threads, by means of which it seizes and draws in the water-insects that chance throws into its neighborhood. This is the only distinctive mark of its animality. In this view, an aerial plant, the *Dionæa* fly-catcher, of which we have already spoken, must be just as much animal as our polyp, as it entraps the insects that venture upon its leaves.

At the bottom of the sea there is a very curious zoöphyte, the *actinia*, or sea-anemone. For a long time it was supposed to be a plant, and esteemed an ocean flower. Whoever has seen in the aquaria of the acclimating gardens of Paris the beautiful, bright-colored *actinia* that tremble on their flexible stems, shaking their colored appendices and the fringes that adorn their heads, can hardly doubt that these charming queens of the water are real flowers. In fact, for centuries sea-anemones were believed to be marine plants.

In the last century coral was supposed to be a marine

---

* A centimetre is a trifle more than one-third of an inch.

shrub, and it was even believed that its flowers had been discovered. The Count of Marsigli, a Paris Academician, won a European reputation by this discovery. Peysonnel, a Provençal naturalist, took great pains to combat the notion, and to show that the supposed flowers were only young corals. Opposed to him was the entire Academy of Sciences; and his obstinacy brought him into such disgrace that he was compelled to quit France and go to the Antilles, there to die in obscurity while practising medicine. All this because he contended that coral was not a plant, and did not produce flowers.

The famous Genevan naturalist, Charles Bonnet, who forwarded by more than a century the science of our time, has given in his book, "Contemplation of Nature," a striking shape to the parallel between plants and animals. We cannot resist the temptation to quote the following passage from this work, in which Bonnet shows so pointedly the difficulty of distinguishing the plant from the animal, and what perplexities attend an attempt to withhold sensibility from the former:—

"Every thing is graduated in Nature: in denying that plants have sensibility, we give a Cornish hug to Nature without rendering a reason for it. We see sensibility gradually diminish between man and the carvel and the mussel, and think that it stops there, that the last-named animals are the very lowest. But there are perhaps many degrees between the sensibility of the mussel and that of the plant: there may be yet more between the most sensible and the least sensible plant. The gradations that we notice everywhere must convince us of the truth of this philosophy: the new degree of beauty that it seems to add to the system of the Universe, and the pleasure found in multiplying feeling creatures, must still more powerfully persuade us to accept it. I will freely avow that this philosophy is very much to my taste. I love to believe that the flowers that adorn our fields and gardens with a beauty ever new, the fruit-trees whose treasures please our eyes and palates, the majestic trunks that compose the vast forests which time seems to have respected, — that all these are feeling creatures which, in their own way, taste the sweets of life. . . . Plants present some facts which would seem to indicate that they have feeling; but I am not sure that we are well situated for observing these facts, or that the firm opinion so long entertained, that they are devoid of feeling, will permit us to judge of them fairly. We ought to have a *tabula rasa* on this question, and to summon the plants to a more impartial and unprejudiced scrutiny. An inhabitant of the moon possessing the same senses and the same fund of intelligence that we have, but not biassed as to the insensibility of plants, would be the philosopher that we seek.

"Imagine such an observer studying the productions of our earth, and that, having devoted some attention to polyps and other insects that are reproduced by cutting, he passes to investigate vegetables. He would, no doubt, take them at their birth. For this purpose he will sow grain of different kinds, and watch its germination. Suppose that some of these grains have been sown in the wrong way, the rootlet turned upward, and the plumule, or little stem, turned down. Suppose, also, that our observer can distinguish the rootlet from the plumule, and knows the functions of each: in a few days he will notice that the rootlet has risen to the surface of the earth, while the plantlet has sunk beneath it. He will not be surprised at this direction so detrimental to the life of the plant, but will attribute it to the position he had given the grains in sowing them. Continuing his observation, he will see the rootlet double on itself in order to get into the earth; and the plumule curve correspondingly in order to rise in the air. The change of direction will seem very strange to him, and he will begin to suspect that the organized creature he is studying is endowed with a sort of discernment. Too wise, nevertheless, to make up an opinion hastily, he will suspend judgment and pursue his researches.

"The plants that our naturalist has been observing sprang up in the neighborhood of a tree. Favored by such vicinity and carefully cultivated, in a short time they have grown much. The soil that surrounds them for some distance possesses two opposite qualities. That at the right of the plants is moist, rich, and spongy; that at the left is dry, hard, and gravelly. Our observer notices that the roots, having begun to stretch out uniformly on all sides, have changed their course, and all turned toward the rich, moist soil. They persevere in that direction until he fears lest they may cut off nourishment from neighboring plants. To guard against this result, he contrives to make a ditch separating the plants that he is watching from those that are threatened with starvation, and in this way he thinks he has made things all right. But these plants that he undertakes thus to direct elude his precautions: they push their roots under the ditch, and reach the other side.

"Surprised at this conduct, he uncovers one of the roots, but without exposing it to the heat: he touches it with a sponge soaked in water; the root soon approaches the sponge. He changes the place of the latter several times, the root follows it, and conforms to all its positions.

"While our philosopher is meditating deeply on these facts, others, quite as remarkable, are disclosed to him almost simultaneously. He notices that all the plants have left shelter and bent forward, as if to expose all parts of their bodies to the kindly rays of the sun. He observes, also, that all the leaves are so adjusted that their upper faces look toward the sun or the open air, while the lower look to their shelter, or to the soil. Some experiments that he has already made have taught him that the upper face of leaves serves mainly as a defence for the lower, and that the latter is mainly intended to suck the moisture that rises from the earth, and to throw off the superfluity. The direction of these leaves, therefore, seems to harmonize fully with his experiments. He becomes more assiduous in studying that part of the plant.

"He remarks that the leaves of some kinds seem to follow the movements of the sun, so that in the morning they are turned toward its rising and in the evening toward its setting. He sees other leaves shut themselves up from the sun in one way, and from the dew in another. He notices a similar phenomenon in some flowers.

"Considering afterward that, whatever may be the position of the plants in relation to the horizon, the direction of the leaves is always nearly such as he had first noticed, it occurs to him to change this direction, and to place the leaves in a position just contrary to their natural one. He had already employed similar means to assure himself of the instinct of animals and to ascertain its capacity. For this purpose, he inclines toward the horizon the plants that were perpendicular, and holds them in that position. By this proceeding, the direction of the leaves is absolutely changed: the upper face, that previously looked toward the sky or the open air, now looks upon the earth or the interior of the plant; and the lower face, that previously looked to the earth or the interior of the plant, now looks up to the sky or to the open air. But speedily these leaves begin to move: they turn on their stipes, as on a pivot, and in a few hours resume their natural position. The stems and branches readjust and arrange themselves perpendicularly to the horizon.

"Every part of a finger-fish, of a mussel or a polyp, is essentially, on a small scale, of the same structure with the whale on a larger scale. It is the same with plants. Our observer, who is not ignorant of this fact, wishes to know if leaves and branches detached from their stock, and dipped in vases of water, retain there the same inclinations that they had on the plant of which they were parts; and experiment proves this to him beyond doubt.

"He puts wet sponges under some leaves, and sees the latter turn toward the sponges and try to adhere to them with their lower faces.

"He observes, moreover, that some plants that he has shut up in his cabinet, and others that he has carried in a case, have turned toward the window of the one, or the air-holes of the other.

"In fine, the phenomena of the sensitive-plant, its various movements, and its instantaneous contraction when touched, constitute the interesting subject of his last researches.

"Laden with so many facts, all of which seem to favor the hypothesis that plants feel, which side will our philosopher espouse? will he yield to these proofs? or will he suspend judgment in the true Pyrrhonic manner? It seems to me that he will adopt the first course." *

---

* Contemplation de la Nature (Œuvres d'Histoire Naturelle de Charles Bonnet, l. viii. pp. 472–484), Neuchâtel, 1781.

Charles Bonnet concludes that the plant as well as the animal possesses sensibility.

According to the system that we have developed, the animal has a soul, as yet imperfect, and furnished only with faculties proportioned to his needs. But since he possesses, in addition to the sensibility that the plant has, intelligence which the plant lacks, we must conclude that the plant is not endowed, like the animal, with a soul properly so called, but only with a rudiment, a beginning, or, in other words, the germ of a soul.

And as we know that the Sun has the power to produce organic life on our globe, his rays promoting the formation of living tissues, plants or zoöphytes, when they fall on the earth or into the water, we must draw from all the foregoing this last conclusion, — that the Sun sends upon the Earth in the form of rays animated germs, which emanate from the spiritualized beings that dwell in the radiant luminary.

Thus our system of Nature is completed; thus, thanks to solar radiation, are connected the two extremities of the vast scale of organized existences whose place and *rôle* in the great theatre of the Universe we have tried to determine. Life begins in the water: it is first manifest in plants and zoöphytes, for these two classes of creatures obey the same laws and seem to have the same origin. The Sun, pouring his vivifying rays upon land and water, promotes in these the formation of plants and zoöphytes, which are the starting points of organization. The animated germ deposited by the

Sun in the plants and zoöphytes grows, in passing from the zoöphyte to the mollusk or vertebrate: it develops still more in the passage from the mollusk or vertebrate to the fish. This germ then becomes a rudimentary soul, furnished with some faculties. It has nothing more than sensibility in the zoöphyte and the mollusk: in the fish, then in the reptile, or the bird, it has the power of attention and judgment. Faculties increase as the animal rises higher in the organic scale. Reaching the summit of this scale, — that is, the human being, — the soul possesses all its faculties, and especially memory, which was obscure and uncertain in the animal stage.

By conceding sensibility to plants, we are enabled to join one with another all the beings of the living creation, and thus to complete our general system of Nature.

CHAPTER XVII.

Proofs of the Plurality of Human Existences and Incarnations. Outside of this Doctrine we can explain neither the Presence of Man on the Earth, nor the Painful and Unequal Conditions of Human Life, nor the Fate of Children who die in Infancy.

THE doctrine of the plurality of existences and of reincarnations, which joins together, like so many links of the same chain, all living creatures, from the most insignificant animal up to those blest beings whose privilege it is to look upon God in his glory; which gives to earthly humanity brothers in the several planets; which makes of the inhabitants of our globe a mere tribe of the Universe; which sees in all the peoples of the worlds but one planetary family, in which every one can rise by his deserts and his struggles in the hierarchy of happiness, — this doctrine is supported by so many proofs, that we are embarrassed only by the necessity of making choice among the means of demonstration which speak in its behalf. To enumerate them all would be to extend this work to an unreasonable length; so we shall content ourselves with bringing into relief some of the most striking of them.

Why are we on the Earth? We have not begged to be, we have expressed no wish to be born. Had we been consulted, we should, no doubt, have preferred either not to come into the world at all, or to come at some other time. Perhaps we should have asked to dwell in some other planet than the Earth. Our globe is, indeed, disagreeable enough as a place of residence. Owing to the inclination of its axis, its climates are wretchedly arranged. One must either succumb to the cold, unless protected against its rigors, or be calcined by heat. In a moral point of view, the conditions of humanity are even more miserable. Evil dominates in the Earth, vice is almost everywhere held in honor, and virtue is so ill-treated that honest living here below is but a certain pledge of misfortune. Our affections bring us only anguish and tears. If we taste for a moment the joys of paternity, of friendship, of love, it is only to see the objects of our tenderness snatched away by death or separated from us by the accidents of a wretched life. The organs given to us for the service of life are dull, gross, and subject to disorders. We are riveted to the soil, and to move our heavy mass costs us fatigue and exertion. If there are a few men well organized, of good constitution and robust health, how many are infirm, idiotic, deaf-mute, blind from birth, maimed, foolish and insane! My brother is handsome and well-shaped: I am ugly, weakly, rickety, and a hunchback. Yet we are sons of the same mother. Some are born into opulence, others into the most dreadful want. Why am I not a prince and a great lord, instead of a poor pilgrim on the earth, ungrateful and rebellious? Why was I born in Europe and at Paris, where by civilization and art life is rendered supportable and easy, instead of seeing the light under the burning skies of the tropics, where, dressed out in a beastly muzzle, a skin black and oily, and locks of wool, I should have been exposed to the double torments of a deadly climate and a barbarous society? Why is not a wretched African negro in my place at Paris, in conditions of comfort? We have, either of us, done nothing to entitle us to our assigned places: we have invited neither this favor nor that disgrace. Why is this unequal distribution of the terrible evils that fall upon some men, and spare others? How have those deserved the partiality of fortune, who live in happy lands, while many of their brothers suffer and weep in other parts of the world?

Some men are endowed with all benefits of mind: others, on the contrary, are devoid of intelligence, penetration, and memory. They stumble at every step in their rough life-paths. Their limited intelligence and their imperfect faculties expose them to all possible mortifications and disasters. They can succeed in nothing, and Fate seems to have chosen them for the constant objects of its most deadly blows. There are beings who, from the moment of their birth to the hour of their death, utter only cries of suffering and despair. What crime have they committed? Why are they on the earth? They have not petitioned to be here; and, if they could, they would have begged that this fatal cup might be taken from their lips. They are here in spite of themselves, against their will. This is so true, that some, in a paroxysm of despair, cut with their own hands their thread of life: with their own hands they pluck out a life which fearful torments render unendurable.

God would be unjust and wicked if he imposed so miserable an existence upon beings who have done nothing to incur it, and have not asked for it. But God is not unjust or wicked: the opposite qualities belong to his perfect essence. Therefore the presence of man on such or such parts of the earth, and the unequal distribution of evil on our globe, must remain unexplained.

Reader, if you know a doctrine, a philosophy, or a religion that solves these difficulties, I will destroy this book, and confess myself vanquished.

If, on the contrary, you admit the plurality of human existences and reincarnation, — that is, the passage of the same soul through several bodies, — all this is made wonderfully clear. Our presence on such or such a part of Earth is no longer the effect of a caprice of Fate, or the result of chance: it is merely a station in the long journey that we make through the world. Before our birth on the Earth, we have already lived, perhaps in the shape of a superior animal, perhaps in that of a man. Our actual life is only the sequel of another, whether we have within us the soul of a superior animal, that we must purify, improve, and ennoble, during our stay on earth; or, having already completed an imperfect and wicked life, are condemned to begin a wholly new one. In the latter case the career of a man begins anew, because his soul is not yet pure enough to rise to the level of the superhuman.

Our sojourn on the Earth is, then, only a kind of probation, imposed on us by Nature, and during which we have to refine our souls, to free them from earthly ties, and faults that weigh them down and keep them from rising, glorious, to the ethereal spheres. Every human life unworthily lived has to begin anew. Thus the student, who has toiled assiduously, rises, at the year's end, to a higher class; but, if he has made no progress in his studies, he continues another year in his old place. Bad men, we think, are those who have lived before, and are going through another year in the same rank. They will be thus delayed till their souls are fit to rise in the hierarchy of beings; that is, to pass, after their death, to the state of the superhuman.

While the cause of our existence here is obscure, and indeed not capable of explanation in common ideas, it is simple and clear in the doctrine of a plurality of existences.

It should be added that this doctrine is conformable to the justice of God. In making earthly life a probation for man, God is equitable and good like the father of a family. Is it not better, indeed, to subject a soul to a probation which, having once failed, it can begin again, than to limit it to a single one, which entails the irremediable damnation of the guilty? It is better to hold out to a degraded creature the possibility of rehabilitation by his own efforts, by his own struggles, than to crush him, all soiled as he is, with imperfections and crimes. The justice and goodness of God are manifest in this paternal scheme, far more than in the severe judgment that irrevocably damns a soul after a single unsuccessful probation.

If human life is a probation, a period in which we prepare for a new and happier existence, we need no longer ask why we are on the earth, why we are living to-day rather than to-morrow, and under this latitude of the globe rather than that; we need no longer ask why we were born into the Earth, and not into Mercury, Saturn, or Mars. Whether we live to-day, or are going to be born hereafter; whether we first saw the light on the Earth, in Mercury, or in Mars; whether we dwell in Europe or Africa, — our destiny remains the same. In fact, we endure a period of preparation that we must accomplish before passing to the superhuman state; and the place, the moment of our journey, the country in which we live, the planet chosen for the theatre of this probation, all these have no bearing on the part we are to play according to the designs of Nature. We have to make a vast journey through the worlds, and a short stay on the Earth is a part of it. In whatever corner of the Universe we may be cast, we shall have to fulfil there the probation that God has set for us, — a probation of suffering and struggles, of physical and moral pain, that we must endure before rising in the hierarchy of creatures. After that, time, place, moral conditions, good or evil, will matter not to us. What we need is a brief sojourn on a planet where this probation can be accomplished; and this may be on the Earth, or in Mars or Mercury, and on any imaginable point of the Earth.

If, in the course of this probation, we encounter moral evil; if we see vice triumphant and virtue prostrate; if we are the innocent victims of the injustice, cruelty, or ignorance of man, — we have not to murmur against Providence, to utter maledictions against pain, to deplore the scandal of crime victorious in the presence of virtue, which suffers and weeps. We have no more to regret the infirmities of our bodies, the maladies that seize upon us in the cradle, and afflict us while life lasts. We have not to lament either the feebleness of our minds or the failure of our faculties. Every condition opposed to earthly happiness makes a part of the programme of probation which we have to undergo here below. Though a thousand ills overwhelm us, though injustice strikes us, though cruel hands fall upon us, we must bless these ills, applaud that iniquity, and kiss those bloody hands. These are the instruments of our natural redemption; and the more piercing and sharp and painful they are, the sooner will come the hour of our delivery, the happy moment of our exit from this impure and sinning globe that we tread for a little while. Besides, justice will soon be done. The wicked man will speedily begin a new life here below, as a punishment for his misdeeds; while the good will rise into the upper realms, where the new life lies before him, broader, happier, wiser, more harmonious with the aspirations of our nature, than the precarious and miserable existence that we drag out here on Earth. There we shall be born again, glorious and strong, with all our memory, all our heart and all our liberty.

Thus vanish the difficulties, thus are solved the problems, thus cease the uncertainties, thus are cleared the mysteries, that no doctrine, no religion, no philosophy, can dissipate, and which lead us almost to doubt the justice of God. The doctrine of reincarnations and anterior existences explains all, answers all.

We pass to one of the most interesting questions of the doctrine of the pre-existence of souls, — that of children dying in infancy. What becomes of the children who die at the age of a few days, in birth, or live eight or ten months? Up to the last-named time of life the human soul has had no development: it is almost in the same rudimentary state as at the hour of birth. What, then, is the fate of these infants after death? Here is the stumbling-block of every religion, and every system of philosophy. And here, on the other hand, as we are going to prove, is the triumph of the doctrine of reincarnations.

The Christian religion is the only one whose opinions on this question it concerns us to know. Let us see what it formulates in this matter, in point of doctrine as well as in point of worship.

The Christian religion affirms that children dying in infancy go to Paradise, if they have received the sacrament of baptism. But this is an arbitrary judgment; for no one can tell what conduct these children would have pursued if their life had followed its regular course. In granting eternal happiness to a soul for an earthly existence of only a few hours, in which the child can have done neither good nor evil, God would be sovereignly unjust. He would be unjust to the rest of mankind, on whom he puts the burden of a whole life of painful trials, in granting a certificate of endless bliss to a being who had spent only a few hours on earth. In order to enjoy eternal happiness, he must have merited it.

This decision of the Church, therefore, cannot be explained, except on the supposition that God is unjust and partial. To create a soul for a life of ten minutes, and to admit it instantly to an eternal reward, is what God could not equitably do.

But to proceed. We have seen what becomes of baptized infants. What happens to unbaptized infants according to the Church? Some fierce theologians, who wrote before St. Augustine, did not hesitate to condemn them to eternal flames. Yet their opinion has not prevailed, and the doctrine of Augustine has become the rule. The Church sends unbaptized infants to a special purgatory, which it calls "Limbo." It is a half-way house between Paradise and Hell. Its inmates are not exposed to endless torment, but they cannot look upon God: it is a medium between the two extremes of eternal rewards and punishments.

This is all very well; but infants who die baptized are very few, compared with the whole human race. The Christian religion is professed by only a little more than one-third of the population of the Earth,* and all Christians do not baptize children. Moreover, many infants die by accident before there has been time to baptize them privately. Certainly five-sixths of the children of men die without receiving the saving sacrament. They go, then, to be lost in the immobility of "Limbo," that cold grave, that somnolent abode of souls which, in their essence, are all activity and motion. God must create, then, feeling souls to cast five-sixths of them into a kind of annihilation!

But this is not all. The institution of baptism is of recent origin: it dates back scarcely eighteen hundred

* The total population of the globe is 1,300,000,000, and the number of Christians is estimated at 380,000,000.

years. Before Christianity all children must have lacked this sacramental ceremony; and hence all, without exception, must have gone straight to "Limbo."

Humanity is very old, — much older than theologians and even philosophers, until within a few years, have believed. Instead of five or six thousand years, humanity goes back perhaps a hundred thousand. So, during a hundred thousand years, infants must have been doomed to "Limbo;" and only within the last eighteen hundred have a few of them, by the grace of baptism, been permitted to enter Paradise. For ninety-eight thousand years all the souls of infants must have gone to people those gloomy cities of the dead. Take notice that all these victims have done nothing to deserve such a fate, for plainly it was not their fault that baptism had not been instituted. So these poor things were punished for a negation, of which they even had no knowledge.

We see of how much account, when analyzed by reason, is the explanation, that is offered by theology as to the fate of children dying in infancy. See now how the same question is simplified in the doctrine of a plurality of existences. It is admitted, in this doctrine, that when a child dies in infancy, — that is, before the age of a soul, which is the established age of dentition, — its soul remains on Earth, and does not pass, like that of grown men, to the state of superhumans. The soul of a twelve-months infant is still in a rudimentary state: it is almost just as it was at the moment of birth. If the infant dies at this age, there is a work to recommence. And the work is recommenced; that is to say, the soul of a child dying before completing its first year, leaving the body with the last breath, goes to take up its residence in the body of another new-born child. After this new incarnation, it begins a second life.

If it happens that the new life lasts no more than a year, there is nothing to prevent the soul from undergoing a third incarnation in the body of a third infant, till it has passed the term that places it within the ordinary conditions.

It is impossible that the soul of a child, as yet undeveloped, and which has added nothing to what it has received, should be treated like improved souls, purified by the exercise of life, and by physical and moral sufferings which make our sojourn on earth a season of preparation and training. The infant could not therefore be admitted to supraterranean regions: it simply begins anew the interrupted probation. The mortality of infants between birth and the age of twelve years is so considerable that Nature must have reserved to herself the means of neutralizing this cause of derangement in the sequence of her operations.*

The explanation here given of the fate of infants, a fate which is the same that awaits bad men after death, is conformable to the economy which is visible in the operations of Nature. Nature will have nothing lost

---

* Statisticians affirm that half of the children born in Europe die before passing the age of one year.

that she has created. The soul of a bad man is evil; but it is a soul: it lives and is eternal; it cannot be lost. Only it must improve and correct itself. This happens, thanks to the new lives to which Nature invites this imperfect soul, to enable it to retrieve itself from its forfeiture. Thus the principle of life is preserved, and nothing is lost of that which was created. No more can the soul of a dead infant perish. A second incarnation in another child will enable it to pursue the course of its development. Thus the soul will be preserved and nothing lost.

Chemistry, since Lavoisier, has brought to light a grand truth, — that nothing of the elements of matter is lost; that bodies change their form, while the material element, the body pure and simple, is imperishable, indestructible, and will always be found in unimpaired integrity in spite of its thousand transformations. If this be true that nothing is lost in the material world, it is equally certain that in the spiritual world there is no loss, only transformation.

Thus nothing is lost in material or immaterial beings; and we may place this new principle of moral philosophy by the side of that principle of chemical philosophy established by the genius of Lavoisier.

CHAPTER XVIII.

The Faculties peculiar to some Children and the Aptitudes and Natural Vocations among Men constitute other Proofs of Reincarnations. Explanation of Phrenology. The "Innate Ideas" of Locke, and Dugald Stewart's "Principle of Causality," are explicable only on the Hypothesis of a Plurality of Lives. Vague Recollections of Former Existences.

IF there are no reincarnations, if our actual life is, as common belief and modern philosophy would have it, an unique fact, which cannot be repeated, our souls must be formed simultaneously with our bodies, and at each birth of a human being a new soul must be created to animate it. We will ask, then, why these souls are not formed on the same type? and why, since all human bodies are alike, there is so great a diversity among souls; that is, in the intellectual and moral faculties that constitute them? We will ask why natural aptitudes are so different and so strongly marked that they often resist all the efforts of education which tries to reform them and to turn them in another direction? Whence proceed, in some children, those precocious instincts of vice or virtue, those innate sentiments of pride or baseness, that sometimes stand out in such striking contrast with the social conditions of families? Why do some children rejoice in the sight of suffering? Why do we see them take a cruel pleasure in tormenting animals, while others are moved, grow pale and tremble at the sight, or even the thought, of a suffering creature? Why, if the souls of all men are cast in the same mould, does not education produce the same effects on all young people? Two brothers are in the same class at the same school: they have the same masters, and the same examples are before them. Yet the one profits wonderfully by the lessons he receives; and his knowledge, education, and manners are irreproachable. On the other hand, his brother remains ignorant and coarse always. If the same seed sown in these two soils has produced such different fruits, is not the reason for this that the soils which received the grain are different?

Natural dispositions and vocations are manifest after the first years of life. There would not be such extreme diversity of aptitudes, if all souls were created after the same pattern. The bodies of animals and of men and the leaves of trees are thus made, for small differences can be detected between those of each class. The skeleton of one man is invariably like that of another man: the heart, the stomach, the veins, the bladder have the same form in all men. It is not so with their souls: they differ materially between man and man. We hear it said every day that one child has a mathematical, another a musical, a third an artistic turn. In others we notice savage, violent, even criminal instincts: after the first years of life these dispositions break out.

When these natural aptitudes are pushed beyond the usual limit, we find famous examples that history has cherished, and that we love to recall. There is Pascal mastering at the age of twelve years the greater part of Plane Geometry without any instruction, and not a figment of Calculus, drawing on the floor of his chamber all the figures in the first book of Euclid, estimating accurately the mathematical relations of them all,—that is, reconstructing for himself a part of descriptive geometry; the herdsman Mangia Melo manipulating figures when five years old as rapidly as a calculating machine; Mozart executing a sonata on the piano-forte with his four years old fingers, and composing an opera at the age of eight; Theresa Milanollo playing the violin at four-years with such pre-eminent skill that Baillot said she must have played before she was born; Rembrandt drawing with masterly power before he could read.

These examples are in the memory of every one. But it should be understood that they do not constitute exceptions. They only hand down for us a general fact, which in these particular cases is extended far enough to attract public notice. They have the advantage of making the common people comprehend an actual law of Nature, by helping them to understand the diversity of aptitudes, and the predominance in some children of special faculties.

There is a phrase in the French language employed to describe children endowed with such precocious and extraordinary talents, who are called "little prodigies." It should be added that this description is sometimes applied in an uncomplimentary sense. Indeed, these "little prodigies" are charged with not holding out as they promised: for this reason it is remarked that these striking aptitudes are by no means guaranties of extraordinary success for any one in his career as a man. The child, who draws with surprising skill at the age of four years, becomes a sad dauber when he has adopted the profession of art. The musician, who ravished his auditors when eight years old, in maturity is but a mediocre performer.

This remark is just, and this explanation should be given of the point. If the "little prodigies" do not become great geniuses, it is because they have not cultivated their aptitudes: they have let their talent become extinct for want of cultivation and exercise. It is not enough to possess a natural inclination for a science or an art: work and study must reinforce and develop this aptitude. "Little prodigies" are often passed in the race by great workers. That is easily understood: they came on the Earth with remarkable powers acquired in an anterior existence, but they have done nothing to develop their aptitudes; they have remained all their lives at the very point where they were at the moment of their birth. The man of genius is he who cultivates and improves incessantly the great natural aptitudes that he brought into the world.

These aptitudes, this predominance of certain faculties noticed in some children, cannot be explained by ordinary philosophy, which requires that a new soul should be created for each child born. They are made very clear, however, in the doctrine of reincarnations; indeed, they are only a corollary of this doctrine. All is plain, if we admit a life before the present one. The individual at his birth brings with him the intuition which results from the knowledge he has acquired during his former existence. Men are more or less advanced in intelligence and morality according to the life they led before coming to Earth to perform the parts in which we see them.

This is evident in the case of a man who begins anew an ill-spent life. He has acquired, in his first existence, powers that are useful to him in his second. He has not, perhaps, in their integrity all the faculties that he had in his past life; but he has what the mathematicians call their *resultant*, and this is special aptitude, vocation. He is an arithmetician, a painter, or musician, because in his first life he had the faculty of calculating, of drawing, or performing on instruments. It is impossible, we think, to find another explanation of our natural aptitudes.

It will be said, no doubt, to be strange that we have aptitudes and faculties resulting from a former existence when we have lost all recollection of it. We answer that one may have forgotten past events, and yet preserve certain powers which are independent of every particular and concrete fact, especially where these powers are very strong. We see daily old men who have lost all recollection of the events of their own lives, who know nothing of the history of their own time or of their own careers, and who yet have preserved their powers, and especially their mere aptitudes. Linnæus, in his old age, loved to read over his own works: forgetting that he was their author, he would say: "How interesting and beautiful this is! I would like to have written it."

There is, then, no evidence against the hypothesis that the child, after reincarnation, preserves the aptitude that he had in his first life, having lost all recollection of past events which he had witnessed during that time. These faculties revive, and it becomes daylight to the child, just as half-extinguished flames are re-excited by a breath of wind. In this case, the breath that makes the obscured human faculties to glow again is that of a second life.

The lack of memory may be urged against the theory of reincarnations in the body of a child; but this argument is worthless when applied to the reincarnation of an animal's soul in a human body. Indeed, the animal being almost destitute of memory, it would seem that only aptitudes can pass from the animal to the man. The instincts, good or bad, gentle or savage, that human souls manifest at so early an age, are accounted for by the kind of animal which has transmitted its soul to the child. The soul of a musical child may have come from the nightingale, the sweet singer of our woods. A child who shows a taste for architecture may have inherited the soul of a beaver, the architect of the woods and waters.

To recapitulate: the different aptitudes, the natural faculties and vocations, are explained without the slightest difficulty, if we accept the doctrine of the transmigration of souls. To reject this, we must accuse God of injustice in granting to some men useful powers that he withholds from others, and in distributing unequally intelligence and morality, the bases of the conduct of life.

This reasoning seems to us unassailable; for it rests not on an hypothesis, but on a fact, — to wit, the diversity of aptitudes, of intelligence and morality, in men. This fact, unexplained in all current philosophic theories, can be explained only in the doctrine of reincarnations; and it forms the basis of our theory.

A good deal has been said, *pro* and *con*, about phrenology; and the result has been general indifference toward it, owing to the apparent impossibility of framing a good theory of it in common ideas. It has been found easier to shut our eyes to the works of Gall than to undertake to explain them. The truth is, Gall was guilty of some errors in detail, as has happened to every founder of a new doctrine, who could not unaided complete a work that had no precedents; but his successors have supplied the deficiencies in his system, and we are compelled now to acknowledge that Gall's theory is exact. It is made up in fact wholly of mere observations, which any one may repeat.

Applied more especially to animals, the theory of Gall, or phrenology, is surprisingly obvious. In the case of man, the facts are almost equally confirmatory of the theory. It is certain that the cranium of the murderer exhibits the abnormal developments marked by Gall, and that, according to the doctrine of the German anatomist, the sentiments of affection, love, cupidity, discernment, &c., can be recognized externally by the protuberances of the bony lamina of the human cranium. It rarely happens that a phrenologist feeling the head of a great criminal, a Papavoine or a Troppman, does not find the dreadful collocation that reveals bad passions and brutality.

Unfortunately phrenology seriously troubles our moralists, whose opinions are weakened by the hackneyed philosophy of our time. The standard moralists ask if a man who bears on his head the bumps of murder is responsible for his crime, if he is personally free, and if he is really so guilty as he is generally esteemed, in obeying the cruel propensities that step-mother Nature has given him? Must we be pitiless to a man who simply obeys his physical organization, almost as a fool obeys the unruliness of his diseased thought? It would seem that it is unjust to punish the murderer; and it is asked if our criminal courts, as well as the scaffold, should not be suppressed, and if the judge who sends to death a man who had no consciousness of his actions is not the real criminal?

The same reasoning, the same uncertainties, array themselves in the case of good actions. Must we be very much pleased with him who is exact in the discharge of his duties, with the conscientious and faithful citizen, with the honorable and good man, if in his wise and upright conduct he merely obeys good impulses which his physical organization had marked out for him in advance?

These deductions from phrenology were, it will readily

be seen, very embarrassing and almost subversive of its morality. The barbarity of society which punishes the culpables, and the lack of honor for the virtuous,—these are sad and painful facts to admit. The rejection of phrenology relieves us from this embarrassment.

It is no longer necessary to reject phrenology: we may retain it, and congratulate ourselves on a new conquest of the science of observation, if we accept the doctrine of anterior existences. Indeed, phrenology is very naturally elucidated in this doctrine. In taking possession of a human body, the soul impresses on the cerebral matter, which is the seat of thought, a modification and a predominance in harmony with the faculties that this soul possessed at birth, and that it has acquired in an anterior existence, human or animal. The brain is kneaded by the soul into conformity with its own aptitudes and its acquired faculties; then the bony envelope of the brain, which moulds itself on the cerebral substance contained in its cavity, reproduces and expresses externally the signs of our predominant faculties. The ancients who said, *corpus cordis opus* (the body is the work of the soul, or the soul makes its body), expressed this same idea with singular conciseness.

We need not, therefore, excuse the murderer; we need not deny him the possession of free will; we need not save him from the just punishment of his crime. It is not because he wears on his head particular bumps that he has dipped his hand in the blood of his victims. These bumps simply reveal externally, as if to warn him, and urge him to correct them, the vicious and evil propensities that he had in himself at birth, and which he should have been able to conquer by the exercise of will, and the ardent desire to reform his base and crooked soul. It is always possible to overcome by sufficient efforts the bad propensities of his nature, and each of us knows how to triumph over pride, indolence, or envy. He who has not known how to correct the vicious dispositions of his soul is culpable, and there can be no excuse for a crime which he has committed in the full enjoyment of his free will. Thus neither society nor God is concerned in this question, if the doctrine of the plurality of existences be admitted.

The English philosopher, Locke, immortalized himself by the discovery that the human understanding has ideas called "innate;" that is, ideas that we bring with us into life. This fact is certain in itself. In our day, a Scotch philosopher, Dugald Stewart, has more precisely stated the discovery of Locke, in showing that the only and true innate idea, that which exists universally in the human mind from birth, is the idea or principle of causality, — a principle which makes us say and think that there is no effect without a cause, which is the beginning of reason. In France, Laromiguière and Damiron have echoed and popularized this discovery of the Scotch philosopher. Thus the standard works of philosophy write down this proposition as a truth beyond all doubts.

We admit unreservedly the principle of causality as

the innate idea *par excellence*, and we shall avail ourselves of the fact. We only ask the ruling philosophy to explain it. Yes: there are innate ideas in our souls, as Locke said; and the principle of causality, which leads us irresistibly to trace effect to cause, is the most palpable of those ideas that seem to make part of ourselves. But why do we have innate ideas? Where did we get them, and how did they come to be in us? This is what the standard philosophy — the philosophy of Descartes, which always rules in France, that is, in the Normal School, and among the professors who have come from the University of Paris — cannot tell us. It will answer, perhaps, employing the favorite argument of Descartes, that we have innate ideas, because such is the will of God, who created our souls. But such an answer would be arbitrary and commonplace: it may be appealed to in any case, and it is not a logical argument.

In the doctrine of the plurality of existences, innate ideas and the principle of causality are very easily explained: they are, indeed, only deductions from that doctrine. Our souls, having already existed, either in the bodies of other men or in those of animals, have preserved traces of the impressions they received during such existences. They have lost, it is true, the memory of actions done during their first incarnations; but the abstract principle of causality being independent of particular facts, and only the general result of the experience of life, must endure in the soul at its second incarnation.

Thus the principle of causality, of which French philosophy can furnish no satisfactory explication, is explained in the simplest manner by the hypothesis of reincarnations and plurality of existences.

We have spoken of memory, and already in another chapter have given our views of it in connection with the subject of reincarnations. We have stated why we are born without any recollection of a former life. If we sprang from an animal, we said, we have no memory, because the animal had none, or only a very feeble one. It should be added that, if we come from a human soul reappearing to the light of day, we are without memory, because memory would have disturbed and even rendered impossible the trial of our earthly life, because it enters into the views of Nature that we should recommence this trial without any trace in our minds of our former acts which would hamper our free will.

We will not leave the subject without remarking that the memory of an anterior existence is not always absolutely wanting in us. Who of us, withdrawn within himself in his hours of solitary musing, has not seen reappear to his gaze a whole world buried in the far folds of a mysterious past? When, absorbed in deep reverie, we give loose rein to our imagination, and it hurries us into the vague infinite, we seem to see magical pictures not wholly strange to our eyes, we seem to hear celestial harmonies which have already charmed our hearts. These secret evocations, these involuntary contemplations, to which each one of us can bear witness, are they not veritable recollections of a life anterior to this?

May we not also attribute to a vague recollection, an unconscious sympathy, the real and profound pleasure we feel in gazing upon plants, flowers, and vegetation? The sight of a forest, of a beautiful prairie, of grassy slopes, touches and moves us, and sometimes melts us to tears. Great masses of verdure, and even the humble field daisy, speak to our hearts. Each of us has his favorite plant, the flower whose fragrance he loves to breathe, or a tree whose emanations and shade he specially affects. Rousseau was moved by the sight of a periwinkle; and Paul de Musset loved willows so tenderly that he made a vow, which was accomplished, that a willow should bend over his grave. This love of vegetation has mysterious roots in our hearts. Must we not see in this natural sentiment a kind of indistinct recollection of a native land, a secret and involuntary evocation of that verdant medium in which the germ of our soul for the first time opened to the light of the Sun, the mighty promoter of life?

Besides the vague and obscure souvenir of visions which seem to belong to our former lives on Earth, we feel, sometimes, lively aspirations towards a gentler and more tranquil destiny than that which we have here below. Probably coarse beings, bound up in their interests and material appetites, do not feel these secret soarings toward an unknown and happier fate; but poetic and tender souls, who endure the sad vicissitudes to which human nature is a martyr and a slave, please themselves with these melancholy aspirations. They catch a glimpse in the glorious infinite of the celestial mansions which one day will be their happy homes, and long to break the bonds that bind them to Earth. Read in Goethe's poem, "Mignon," the touching episode in which Mignon, a wanderer and exile, poured out her young soul in longings for heaven, in sublime soarings toward an unknown and blessed future, which she knew by presentiment and which lured her toward itself, and ask yourself if the beautiful verses of the great poet, who was a great naturalist also, do not reveal a truth of Nature,—the new life that awaits us in the plains of ether.

Why, among all men and all peoples, are the eyes turned to heaven in solemn moments, in bursts of passion, in the anguish of pain? Was any one ever known at such times to gaze with the same intensity upon the ground, or whatever stretches beneath his feet? It is always to heaven that our eyes and our hearts turn. The dying bend their failing sight heavenward; and toward the celestial realms we look longingly, when rapt in one of those vague reveries that we have just described. We may believe that this universal tendency to look toward heaven is an intuition of what lies beyond our earthly life, a natural revelation of the domain that will be ours some day, and which stretches into the celestial realms, to the very bosom of ether.

CHAPTER XIX.

The Hypothesis of Successive Existences compared with Materialism, and with the Dogmatic Christian View of the Destiny of Man. Punishments and Rewards in the Christian Scheme, and in the Doctrine of Successive Existences.

PROPERLY to appreciate the hypothesis of successive existences and reincarnations, we must compare it with the idea of the destiny of man fostered by the principal philosophic and religious systems. We cannot enter into an exhaustive examination of all the philosophic or religious conceptions that have held place in the history of the human mind; but we shall give a summary of such opinions, in addressing ourselves to materialism on the one side, and to the doctrine of Christianity on the other, to inquire how they explain the origin and end of humanity.

Materialism means that the sensible principle that we have within us is wedded to the body, and that at the instant of death it is destroyed, like the body itself. It is a torch that is extinguished, and will never relume itself. "Man," say the materialists, "lives and dies like plants and animals. Born of a germ, like plants and animals, like them he is developed. On the death of the body, the sensible principle, which is merely the result of organization, is extinguished with the organs, and never lives again. We see every thing around us perish, and nothing reappear. Why should man be an exception to the general rule?"

We will see to what consequences we are led by this system, purely negative as it is, born not of science, as has been too often said, but of ignorance and disregard of all the facts of Nature.

A man has spent his whole life in crime; he has trampled under foot every thing just and good; he has crushed the weak and oppressed the innocent: his whole life has been a prolonged insult to humanity. In spite of his crimes and his pollution, he has known how to retain the esteem of his fellows, whom he is skilful in deceiving. When he dies, he passes away quietly, and with a serene heart. Is this great criminal going to meet, after death, the same destiny with his miserable victims; and is the putting of his foot within the tomb enough to save him from all punishment, all expiation? There have been Attilas who marked their paths along the Earth with rivers of blood and heaps of ruins; there have been Yenghis-Khans, who have ravaged vast tracts of country with fire and sword, and have bathed in the blood of thousands of men. Was all done with them when they went down to the grave? Did their innumerable victims do nothing else, after their death, but share the lot of their executioners?

Look, on the other hand, at a man who has sacrificed his life to the performance of obscure duty. He has given his blood or his powers to the service of his country or of humanity. In reward for a lifetime of toil and devotion, he has reaped nothing but indifference, misery, and disdain. He has lived, humble and modest, with no wealth beyond a day's wages at his plough. When this good man shall have returned his pure and holy soul to God, will all be ended with him? Will death be as bitter as life was to him? and can he not hope for, after this life, another destiny than that of the great criminals who were the terror of mankind?

On the burning shores of Africa, on the marshy banks of the Indus and the Ganges, there are whole peoples whose lot is more wretched than that of animals. Slaves or Pariahs, they are subjected to the caprices and the brutal will of a master. They are sold like poor cattle: their children do not belong to them, and they belong not to themselves. Will you have it that these miserable creatures do not find beyond the grave some compensation for the direful evils of their lives? Will you have slave and master share a common fate? Will you have it that he who has tortured the Pariah, who has doomed his days to pain and weariness and abject misery, is not distinguished from his hapless victim? Will you have it that both fall into the gulf of annihilation? Our reason and our hearts protest against such an idea.

If it were thus, the moral order, the harmony, which we apprehend between desert and recompense, between crime and punishment, would be utterly destroyed. While, then, regularity and equilibrium are seen in Nature, in what concerns man there would be a universal disarray, — real sixes and sevens. Of what use would it be to be honest, good, faithful to obligations, devoted to duty? We should have to arm ourselves against each other to wage wars of extermination and endless quarrels; to seek in force, violence, and all bad passions, the means of triumphing over our neighbor and of securing for ourselves the greatest amount of brutal pleasures.

One more consideration. We lose friends, parents, brothers, sons. Our separation, painful as it is, must then be final. Nothing will be left of those beings who have made our hearts throb with the tenderest emotions. You have a mother, the joy of your grateful soul; and when death robs you of her, you must be told that you will never see her again, — that you are forbidden to hope again to look upon her, in that inevitable day when death shall come to make you obey, in your turn, the common law of Nature. You will be doomed to sink, like your mother, your son, your friends, into an eternal sleep; and your only consolation will be the thought that you all fall into the same abyss of destruction and annihilation.

It is easily explained, however, how on the Earth the moral order and the natural equilibrium can be often disturbed. Human wisdom is limited and subject to weaknesses, social conditions are sometimes tyrannical, and we cannot hope to realize that ideal equity that we conceive of, and which belongs to perfect beings. But these shocking irregularities cannot touch Providence, which is the principle of all order, and the ideal of all good. We cannot believe that it comes within the designs of God to let virtue be always prostrate, and vice triumph unpunished. We cannot believe that God is so imperfect. Such a God would be no better than we are.

But to contend that God is imperfect and faulty is to deny his existence; for in His nature God is sovereignly just, sovereignly perfect, and to deny Him these attributes is to deny His existence.

Thus the system of the materialists is atheism pure and simple. But the existence of God is incontestable; for he is merely the Supreme Cause of the effects that we witness, and every effect implies a cause. It is plain, in this view, that materialism leads to atheism; and that, moreover, it is false reasoning.

All creatures are conscious here below of desires which do not transcend the assigned limits of their destiny. The animal feels needs: all his needs are satisfied; he asks nothing more; he cherishes no aspiration that is not realized. It is not so with man. In the depths of his soul he feels infinite longings for happiness, for expansion by spiritual growth. He comprehends the meaning of perfect justice, and longs to see it rule about him. He has noble desires, and would like to see every thing harmonize with the ideal sentiment of truth and justice that lives within himself. He would like to soar beyond the narrow bounds that confine him, and cross with rapid flight the vast expanse of the heavens. None of his desires is ever satisfied. If beings inferior to man realize all their longings, shall man, who so greatly surpasses them, be denied this blessing?

It is beyond doubt that this ideal thirst for justice that we feel within ourselves will some day be satisfied, that the equitable distribution of good and evil not vouchsafed to us in this life may be realized after death. Punishment for sin and reward for virtue await us beyond the tomb: this is what the feeling within us proclaims. Materialism, that preaches annihilation, and scouts the idea of rewards and punishments after death, contradicts the strongest of our inner sentiments.

The simple fact of our existence proves, it seems to us, that we shall not perish. God has placed us in this world. He has made us men and women. He was free to create or not to create us; but from the moment when He called us to be here, He could not annihilate us; for that would be, in the beautiful phrase of Malebranche, to show inconstancy, and God is unchangeable in His plans. He builds not to pull down afterwards; He raises not to overthrow. He gave us life because such was His will; and His will would be only a caprice, if after having given us life He robbed us of it or withdrew it. He cannot repent of His works, and take from us that which He has given. If this were possible, He would be made in our image, with our passions and errors: in other words, HE would not be.

Therefore simply because he is, man is immortal. His body vanishes; but his body is nothing. The soul is every thing; and the *being*, the human being, is eternal.

Many other reasons could be adduced to show that materialism has no philosophic basis. We will not prolong this course of argument, whose sole end is to establish that the hypothesis of successive existences is, in all respects, far superior to that dreadful system, which is only a negation pure and simple, and which has therefore no merit, scientifically or philosophically viewed. To deny is easy; but it is not a philosophy: it is a question of explanation. We are surrounded not with negations, but with facts, realities: these facts and realities must be taken account of; and materialism, which explains nothing, which slinks away and excuses itself, is not worthy of being considered as a philosophic conception.

If materialism is the scourge of society, religion is its savior; if the one desolates the heart, the other soothes and strengthens it. Materialism has never troubled itself about the want of morality in its principle touching rewards and punishments hereafter. On the other hand, all religions are much concerned with this question: the law of rewards and punishments after death has a prominent place in them all, and especially in the Christian religion.

Unfortunately the tenet of rewards and punishments in Christianity, conceived more than two thousand years before Jesus Christ, bears marks of the ignorance of those remote ages. It makes God in the image of man: it invests the Creator of the Universe with our passions, our paltry and limited justice. Springing up before the birth of astronomy, strengthened by nothing but the deceitful reports of the naked eye and vulgar errors, it saw only the Earth, and took no account of all the rest of the Universe.

The Fathers of the Church, who, after Jesus Christ, were called upon to give a definite form to religious tenets, scrupulously respected them. The true mechanism of the world was then unknown; for no astronomical system except Ptolemy's had any honor. Saints Augustine and Jerome, unable to grasp the grandeur of the Universe, gave no thought to astronomy, and strove only to revive some minor parts of the Biblical system.

Thus moderns who profess Christianity, or one of its many derivatives, find themselves holding to-day, as to the destiny of man after death, the same childish opinions that the imagination of the Orientals conceived four thousand years ago, in a period of universal ignorance and social barbarism. At all events, the Catholic Church puts the doctrine of future rewards and punishments in this way: After death our body remains on Earth, and there undergoes the decomposition which destroys its material structure. Our soul appears before the bar of God, who judges it sovereignly. The souls of the good repair to Paradise, where they are to enjoy eternal bliss. The souls of reprobates descend into hell to endure endless torments. The souls of those who have not sinned mortally are remitted to Purgatory, where they are held, as it were, under advisement, and whence they may be released by the intercession and prayers of the holy.

The bodies of all men remain on Earth after death; but they will not remain here for ever. At the last day the trump of God's angel will resound throughout the world. At its voice every tomb will open. Bodies will resume their former shapes, and the souls that had abandoned them will return to possession. Not till then will be determined the definitive destiny of man. Reclothed in their bodies, the elect of God will dwell eternally in Paradise, where they will sing His endless praises; while the damned, hurled into the gloomy Gehenna of hell, will there endure all the anguish and torment of never-ending pangs.

The features of Grecian and Roman mythology are very evident in this picture. The Christian Paradise and the Champs Elysées are all one, as hell is the same in Christianity and mythology. Modern religions other than the Christian have also their Paradise and their hell, copied from Greek and Roman antiquity. The Paradise of the Mahometan is even more human and gay than the Christian's. In all modern religions, as in the ancient mythology, we find God set up in the place of a judge, who pronounces the doom of men after death, and rewards them by torment or perennial bliss, in hell or in Paradise.

These childish conceptions, born in the infancy of the world, are merely poetic legends, pleasing or frightful. Of course, we shall not undertake seriously to refute them. The Christian tenet as to rewards and punishments is a dream of the Oriental fancy, and it would be folly to crush it with logical argument.

Other writers enough have taken this trouble. After Diderot, Voltaire, and the Encyclopædists of the eighteenth century, there is nothing to glean in the beaten field of scepticism; and it would do no good to rejuvenate their demonstrations, or warm up their stale sarcasms. Besides, it would pain us to subject to cruel dissection opinions that still have power to comfort souls; that are held by honorable, noble, virtuous, and sincere men; that are, even to-day, the only barrier that we have to oppose to the hateful principles of materialism; and that, in fine, have a most affecting and laudable end, — to guide men in the paths of duty, of virtue, and of hope in another life.

Let us not forget to add that the Catholic Church has had the good judgment to put every one at ease. She has anticipated ideas which she saw to be too feeble to endure the test of criticism. The Church declares that its tenets generally are articles of faith. Now to have faith is to shut the eyes of the mind, — to believe in dogmas, notwithstanding the evidence of the senses and the reason. With this ingenious and convenient purpose of non-reception, which cuts off all question, the Church could add: *Credo quia absurdum* ("I believe it because it is absurd"). One could not put a better grace on the confession of the weakness of his cause.

Let us, then, leave out of the question, or regard as mere myths of the Oriental imagination, these human bodies sunk in putrefaction, vanished, turned to dust, or burned, which in the day of final judgment stand still intact and ready to receive their souls, which come from hell or Paradise to take on their former material garments, and to return, muffled in their old bodies, to the realm of bliss or of eternal torments; this last judgment fixed for the last day of the world, which, very probably, whether we speak of the Earth alone or of all the planetary worlds, will never have an end; these souls which, in hell, or awaiting the last judgment, are subjected to the direst suffering, although, being deprived of their bodies (pure spirits), they cannot suffer; these torments inflicted upon human beings, unnecessary and purposeless, since they cannot lead the sinner to repentance or back to virtue, since there is nothing at the end of this fearful expiation, since forgiveness is not to follow these atrocious severities, — the damned having to be always and endlessly tortured, with no other result than their blasphemies and anguish; that crying injustice which inflicts a punishment infinitely long for a momentary fault, for only one ill-spent lifetime, sometimes even for an unwitting error; that sleepy Paradise, where souls, ranged on benches, do nothing but gaze on the glory of God and chant His praises, where eternal immobility is the law, while the true law of beings is motion, incessant activity, a continual pushing toward progress, elevation by labor, — the labor which is the law of Nature and the very essence of God, and which must be also the rule, the law, the principle, of souls arrived in the heavenly mansions; that judgment, as of the merest trifle, of eternity, which dooms you, with the stroke of a pen, to endless bliss or torment, — as if eternity were an element that the human mind could not only meet with, but even comprehend, — as if eternity were not a vast gulf in which reason loses itself, — as if man could pass, in imagination, the limits of the finite, — as if it were not enough to impose a punishment of unknown duration, or incalculably long, or both, still proportioned to the sin, without opening the useless perspective of that infinity before which the human mind recoils in terror, when it is bold enough to peer over this abyss of mysteries; this God made in the image of man, clothed with the evil passions of humanity, — cruel, vindictive, jealous, now angry, now appeased, as if God had any feeling like those in our feeble hearts, — as if evil were not the exclusive inheritance of human impotence and insignificance, — as if there could be evil in God, who is all-powerful because He is all order and harmony, — as if all the evil on Earth did not proceed from nothing but man's abuse of his freedom; and, finally, that strange dogma in which, of the whole Universe, with its innumerable worlds, we see only the Earth, know only the Earth and its inhabitants, — the Earth, a paltry atom, lost in immensity, a grain of dust compared with the millions of globes with which space is filled!

Compare these ideas with the system of a plurality of existences and reincarnations, and answer if the latter is not the most satisfactory, not only to the mind, but to the heart; if, besides being in harmony with scientific information as to the multiplicity of planetary worlds, the doctrine of successive existences has not the further merit of conformity to justice, to equity, and to morality, that is, to the idea that we have formed of God.

In this doctrine, Earth is not the end of the calculation. Our life on this globe is but the sequel of another life; and what we have achieved in an earthly life we shall achieve in the life following, be it on the same globe, or in the domain of ether. Our actual life being only a period in which we ought to improve, purify, and ennoble our souls, after this probation we shall be dealt with according to our actions and our deserts. Criminal and wicked men, coarse and vile souls, will recommence their lives here below: such is their punishment, and their opportunity granted by Nature to rise from their degradation. Good and feeling men, with souls exalted and refined by the practice of virtue, will quit this imperfect globe, and in the form of superhuman beings will enter upon the plains of ether, with individuality, conscience, memory, and freedom unimpaired. The knowledge that man has acquired during his first life will remain his property in the life following. He will enter the ethereal realms just as he shall have left his earthly residence, with the faculties of his soul just as they were at the moment of his death. As Charles Bonnet said, "The progress that we shall have made here below in knowledge and virtue will determine our point of departure in another life, and the place we shall occupy there." *

* Philosophic Palingenesis.

The wise and the ignorant will not stand on the same footing in the new life: they will be separated by the diversity of moral power which will result from the inequality of their mental stores, acquired during earthly life; and, as Charles Bonnet says, they will divide at this point, to complete their new careers.

A man has passed his life bent over books. Having stored his mind with various knowledge, he has opened new views in the study of human nature, he has added to the power of the human mind. Would you have this man, after death, enter upon exactly the same destiny as the ignorant, brutal, and degraded being, who has acquired nothing, learned nothing, and has suffered his own soul to degenerate? This cannot be. The knowledge acquired by any one is a good that cannot perish, that is sure to be met with again somewhere. Nature wastes nothing. A force once created is not annihilated: it always reappears. The vast amount of knowledge gained by the wise man of whom we spoke must therefore be directly profitable to him after his death.

This leads to the remark that the good man and the guilty will never share the same celestial promotion. The executioner and his victim will not take each other by the hand in the sublime realms. Yet the wicked will not be for ever shut out from the Eden reserved for spotless souls; he will not see the gate of heaven closed upon him for ever. He will pass it, and enter the kingdom of peace and supreme happiness, when the fruits of his earthly life shall have made him worthy of such exaltation.

All this, it should be understood, bears the stamp of morality and justice. Is it not better, indeed, that man, however vicious, however degraded he may be, should continue in his individuality to cherish the hope of a saving reformation; that he should be permitted to raise himself, by his return to virtue, in the hierarchy of beings, than that he should be irremediably condemned, after a single unsuccessful probation, after only one ill-spent life? It is the part of a just and good God to grant the sinner a chance of success in an effort which may save his life, which would be lost without it. God builds not to tear down immediately: He does not permit intelligent souls, that He has created in His sovereign power, to incur annihilation. He leaves them the possibility of retrieving their fall, and of returning to the bosom of Nature, within the circle of the activity of life. He does not despair of his work. A good workman does not cast aside as a failure an outline that has come from his hands: he takes it back, and finishes it. So God takes back and improves upon a first unsuccessful attempt. He wills that no power shall be lost, that none of His creatures shall remain useless and barren toward himself and toward Him.

We cannot subject to critical examination the Christian tenet of punishments and rewards: we have contented ourselves with stating its terms precisely, to show it in parallel with the doctrine of the plurality of existences.

Yet we cannot refrain from remarking, in closing

this chapter, how far less consoling is the Christian tenet than the doctrine that we are explaining. If this tenet were the expression of the truth, the ties of our affections would be broken in a manner often cruel and irreparable. We have sons, brothers, friends, who are as dear to us as ourselves, and who, so to speak, live in our souls. The judgment of God, which operates after a single earthly life, — a life in which the chances of evil are so many and so fatal, — hazards the separation of two persons who are bound together by the ties of kindred or the closest friendship. One of them, if he has incurred the wrath of God, may be plunged into the infernal gulf; while the other, in reward of his virtue, may be called to the eternal bliss of Paradise. Behold the father and the son, the wife and husband, two dear friends, doomed to opposite fates! Behold them separated for eternity! Amid the perfect happiness of the home of the blest, the father will be tortured by the agonizing thought that the son, whom he has loved so dearly, is for ever parted from him, — that that being, the object of so great solicitude and fervent love, is condemned to an eternity of suffering, to torments without end. Thus the sentiments of affection, which would have made the happiness of those two beings here below, will in the upper world make their undying despair. Both of these would be conscious of the sentiment of paternal tenderness or mutual friendship, only to regret its irrevocable annihilation.

This absurd inconsequence cannot be imputed to the doctrine of successive existences. While the Christian tenet threatens to separate from us the objects of our affection, to condemn to eternal separation the souls that loved each other on earth, the doctrine of a plurality of lives simply postpones the moment of their reunion. If one of them is delayed a little in his exile on Earth, by reason of lapses or errors in his life, he can retrieve himself in the life following, and soon rejoin the soul that awaits him in the celestial realms. The reunion always takes place at last, and is delayed only in order to give the imperfect and fallen soul time to render itself worthy of the one that loves it, and of becoming its equal in spirit and perfections.

Thus the plurality of lives gives us the assurance that, whatever may happen, we shall one day be re-united to those we love. It even tells us that this reunion will be speedy, will take place immediately after death, if both parties have spent their lives in conformity with the general laws of the moral order.

This doctrine, then, holds out the surest and most soothing consolations to those who have been plunged in grief by the loss of a dear friend. It is the sovereign balm for wounded hearts. We know that those whom we love, and whom death has snatched from us, are not lost. We see them through the kindly light of this comforting doctrine. They are only hidden for a moment from our weeping eyes. Soon we shall see them again amid the dazzling light in whose streams bathe the worlds on high. We know that they wait for us on the threshold of that shining home, and that they will prepare the way to those sublime abodes that will be the reward of our virtues, the recompense for our sufferings bravely endured, as they have been the payment and the crown of the virtues of those for whom we grieve. We are assured that we shall spend with them the infinite life that unfolds itself for us through space; and we know that this happy reunion with those whom we have loved, and whom we love always, will not be thwarted by any accidents, or hindered by any of those obstacles that checked and saddened here below the mutual outpourings of our hearts. Is there a doctrine sweeter and more comforting to afflicted souls?

CHAPTER XX.

Summing up of the System of the Plurality of Existences.

WE think we ought to recapitulate, in a few summary propositions, the principal features of the system of Nature that we have been expounding.

I. The Sun is the primary agent of life and organization.

II. In the primitive days of our globe, life first appeared in aquatic and aerial plants, and then in zoöphytes. The same order is maintained even to-day in the beginning and development of life and of souls. The solar rays, falling on the earth and the water, promote the formation of plants and zoöphytes. By depositing in these media animated germs, emanating from the spiritualized beings who dwell in the Sun, the solar rays stimulate the birth of plants and zoöphytes.

III. Plants and zoöphytes are endowed with life and sensibility. They enfold an animated germ, as the seed enfolds the embryo.

IV. The animated germ contained in the plant and zoöphyte passes, at the death of these, into the body of an animal, which next succeeds it in the ascending scale of organic growth. From the zoöphyte the animated germ goes into the mollusk, then into the vertebrate, the fish, or the reptile. From the body of the reptile it passes into that of a bird, and finally into that of a mammiferous animal.

V. In traversing the entire series of animals, this rudimentary soul improves itself, and acquires the beginnings of faculties. To feeling, conscience joins itself, will and judgment. When the soul reaches the body of a mammifer, it has acquired a certain number of faculties. Besides feeling, it has the basis of reason; that is, the "principle of causality." From a mammifer of the superior orders the soul goes into the body of a new-born infant.

VI. The infant is born without memory, as was the animal from which he came. He acquires this faculty when about a year old, and enriches it gradually with new powers: imagination and thought are developed; reason strengthens itself; memory grows firm and expands.

VII. If the child dies before the age of about twelve months, his soul, yet very imperfect, and provided with active powers, goes into the body of another new-born child, and begins a new life.

VIII. At the death of a man, his body remaining on Earth, his soul rises through the atmosphere as far as the ether that surrounds all the planets, and enters the body of an angel, or a superhuman being.

IX. If, during its sojourn on this Earth, the human soul has not been sufficiently purified and ennobled, it begins a second life on the Earth, passing into the body of a new-born child, and losing the memory of its former existence. Only when it has attained the requisite degree of moral improvement can this soul, after being reincarnated once or twice, quit our globe, to take on a new body in the depths of ether, and become a superhuman being, who regains the memory of his anterior lives.

X. What happens on the Earth happens also on the other planets of our solar system. In Mercury, Venus, Mars, Jupiter, Saturn, Uranus, and other planets, the same operations take place. In the planets, the Sun provokes the birth of vegetables, or things analogous to our vegetables. By the influence of its rays falling on these globes, and discharging thereon its animated germs, it produces plants and the inferior animals. Then these animated germs contained in plants and inferior animals, passing successively through the whole series of animals, at last generate a being superior by intelligence and sensibility to all other living beings. This being, the analogue of the human, we call planetary man.

XI. When the planetary man, resident in Mercury, Venus, Mars, Jupiter, or Saturn, dies, his material envelope remains on the planetary globe; and his soul, if it has become suitably purified, passes into the ether that surrounds each planet. It goes to be incarnated, and produce a superhuman being.

XII. The phalanxes of superhuman beings float, then, in planetary ether. There is the reunion of all purified souls come from Earth, and from the other planets. The organic type of these beings is, moreover, the same, whatever may have been their planetary country.

XIII. The superhuman being has special attributes: he is endowed with powerful faculties, which elevate him infinitely above terrestrial or planetary man. In this being, matter is reduced, in relation to spiritual principle, to a proportion much smaller than in man. His body is light and vaporous. He has senses which are unknown to us; and those that we have are in him marvellously improved, enlarged, and subtilized. He can transport himself in a brief time to any distance, and travel without fatigue. His sight covers immeasurable extent: he has an intuitive knowledge of most of the facts of Nature which an impenetrable veil conceals from feeble humans.

XIV. The Earth-born superhuman can put himself in communication with men who are worthy to hold communion with him. He directs their conduct, watches over their actions, lights their reason, inspires their hearts. When they come, in their turn, to the heavenly home, he receives them at the threshold of these new realms, and teaches them to live the happy life that awaits them beyond the grave.

XV. The superhuman is mortal. When in the bosom of space he has completed the natural course of his life, he dies, and his spiritual principle enters into a new body, the archangel or the arch-human, in whom the proportion of spiritual principle predominates more and more, compared with the component matter.

XVI. These reincarnations, in the very depths of ethereal space, are repeated, — just how many times it is impossible to determine, — and produce a series of creatures more and more active in thought and powerful in action. At each promotion in the lofty hierarchies of space, these sublime beings see increased the energy of their intellectual and moral faculties, their power of feeling, their power of loving, and their knowledge of the profoundest mysteries of the Universe.

XVII. When he has reached the last degree of the celestial hierarchy, the spiritualized being is absolutely perfect in power and intelligence. He is then wholly divested of all material alloy. He no longer has a body: he is pure spirit. In this state he attains the Sun.

XVIII. The Sun, the sovereign star, is, then, the final and common home of all spiritualized beings, come from different planets, after passing through the long series of existences which have glided away amid the boundless plains of ether.

XIX. These spiritualized beings, united in the Sun, transmit to the Earth and the planets emanations of their essence,—that is, animated germs,—which distribute on the planets organization, feeling, and life, at the same time directing the great physical and mechanical processes which go on upon the Earth and in the other planets of our solar world.

XX. The formation of aquatic and aerial plants, and the birth of inferior animals or zoöphytes, is, we have said, the result of the action of solar rays on our globe. Then begins the serial transmigration of souls through the bodies of different animals, which ends in man, in the superhuman, in the whole girdle of celestial metempsychoses, the last term of which is the spiritualized being, or the inhabitant of the Sun.

Thus is closed and completed this grand circle of Nature, this unbroken chain of vital activity, which has neither beginning nor end, and which binds together all beings in one family, the universal family of the worlds.

Nature, then, is not a right line, but a circle; and we cannot say where it begins or where it ends. Egyptian wisdom, that represented the world by a serpent coiled upon itself, was the symbol of a grand truth, which is going to bring back to light the science of our day.

CHAPTER XXI.

Answer to Some Objections: 1. *The Immortality of the Soul, which forms the Basis of this System, is not demonstrated;* 2. *We have no Recollection of Former Existences;* 3. *This System is only the Metempsychosis of the Ancients;* 4. *This System is confounded with Darwinism;* 5. *This System is borrowed from Fontenelle's "Plurality of Worlds."*

AFTER this recapitulation, which brings into relief the whole doctrine of successive existences and reincarnations, we think we ought to meet some objections that the enunciation of these propositions has provoked, and give them an answer which will serve to complete, on several points, the exposition of our ideas.

First Objection. — It will be said at first: *The existence of an immortal soul in man is the basis of all reasoning. Now the fact of the existence of an immortal soul is not demonstrated in the course of the work, and cannot elsewhere be demonstrated.*

Here is our answer to this first objection.

We are composed of two elements, or substances: one that thinks, which is the soul, or the immaterial substance; and another that does not think, which is the body, or the material substance. This truth is self-evident. Thought is a fact certain in itself; and it is another fact, equally certain, that my arms, my nails, or my beard, do not think. Here, now, is proof of the immortality of the thinking principle, or soul.

Matter does not perish. Observation and science show, in fact, that material bodies are never annihilated, that they merely change their states, form, and place, but are always found again somewhere, substantially intact. Our body is decomposed, dissolved; but the matter of which it is formed is never destroyed. It is lost in the air, in the earth, and the water; it produces in these media new material combinations: but it is, nevertheless, never destroyed. Now if matter does not perish, but is merely transformed, *a fortiori* the soul must be imperishable and indestructible: like matter, it undergoes changes, but never destruction.

Descartes said, "I think, therefore I am." This reasoning, so much admired in the schools, always seemed to us a piece of simplicity. He should have said, to give his syllogism force: "I think, therefore I am immortal." My soul is immortal, because it exists; and it does exist, because I think.

Therefore the fact of the immortality of the spiritual principle within us is self-evident; and to prove the existence of the soul we need none of those demonstrations that fill works on philosophy, and which from antiquity down to our day, from Timæus of Locris to M. Cousin, have paid the expenses of Treatises on the Soul. The difficulty is not to prove that there is a spiritual principle in us, — that is, one that resists death; for to question the existence of this principle we must doubt thought. The true problem is to ascertain if the spiritual and immortal principle within us is going to live again after death, in ourselves or somebody else. The question is, Will the immortal soul be born again in the same individual, physically transformed, into the same person, — into the *me*, according to the felicitous phrase of French philosophers, — or go to make part of another being, a stranger to itself?

Now observe that the only interest for us of this question is just here. It would matter very little to us, really, whether the soul were immortal or not, if, being actually indestructible and immortal, it were going to serve some other than ourselves; or even if, coming back to us, it did not preserve the memory of our former life. The resurrection of the soul, without any recollection of the past, would be veritable annihilation, — the nothingness of the materialists. It must be, therefore, that the soul lives again in our very selves, after our death, and that this soul has then the memory of what it did in its anterior existences.

In fine, it is a question of knowing, not whether our soul is immortal, — that fact is self-evident, — but if this soul will be kept for us in another life, if we shall have after death identity, individuality, *personality*. That is the whole case. Now it is precisely to the study of this question that this work is devoted. We seek to prove here that the soul of man remains always the same, notwithstanding its numerous peregrinations, despite the changes of form in the bodies which it successively inhabits, when it passes from the animal to the man, from the man to the superhuman being, and from the superhuman, after other celestial transmigrations, to the spiritualized being who lives in the Sun.

We endeavor to prove that the soul, in spite of its journeys, in the midst of its incarnations and divers metamorphoses, remains always identical with itself; only, at each metempsychosis, each metamorphosis of the external being, improving and purifying itself, growing in power and intellectual grasp. We try to prove that, maugre the shades of death, our individuality is never destroyed, and that we are born again in the heavens with the same moral person that we have here below; in other words, that the human personality is imperishable. It is for the reader to decide if we have accomplished our purpose, if we have established the truth of this doctrine according to the rules of argument and the facts of science.

If an absolute demonstration of the existence of an immaterial principle be insisted on, we answer that philosophy, like geometry, has its axioms, — that is, its self-evident truths, — which may not, or, if this form is preferred, which cannot, be mathematically demonstrated. The existence of the soul is one of these axioms of philosophy. To a disputant who denied motion, Diogones made answer by walking before him. In the expression of a thought, in saying "Yes" or "No," must be proved the existence of the immortal soul to the sophists who have presumed to question it.

We just said that geometry has its axioms. Well, we must know that a whole school of geometricians, lovers of piquant sophisms, amuse themselves with discussing axioms under this pretext, that it is impossible to prove them. We were present, in December, 1869, at a curious *séance* of the Institute, in which M. Lionville, the famous mathematician, and professor at Sorbonne, expounded this strange piece of polemics with great subtlety.

When we undertake to demonstrate the propositions of geometry, we necessarily begin by admitting certain axioms; that is, some self-evident truths. Else on what basis should we build the first reasoning? But of several propositions of this kind which address the thought, and one of which being admitted the others flow from it, which is the most evident? That depends on the nature of the mind of each of us; and for this reason there never has been, and never will be, harmony on this question.

There is one school of geometry which undertakes to demonstrate every thing. There is another — and this is the good one — which, realizing that the human mind has limits, and that all things cannot be reached by our thought, gives the name of axioms to some truths which need no demonstration, or which are not susceptible of demonstration, which is often the same thing.

Among self-evident truths, or truths very difficult of demonstration, is the question of parallel lines. What are two parallel lines? Two lines that never meet. But how can this property of these two lines be proved by reasoning? That is not strictly possible, since the idea of infinity is not admitted, or not comprehended by every one, and therefore cannot serve as the basis of absolutely exact ratiocination.

For this reason, Euclid, the founder of geometry among the ancients, laid down this truth as a single axiom, and demanded (whence comes "Euclid's postulate," from the Latin *postulare*, to demand) that the truth of this principle should be granted, knowing his inability to prove it by logical demonstration.

A hundred geometricians since Euclid have attempted to demonstrate this theory of the parallels, which he declined to demonstrate; but none have succeeded. In connection with a new attempt at demonstration, made by M. Carton, a mathematician of the Provinces, M. Lionville had addressed the Academy, for the purpose of setting forth the opinions on this point which geometricians almost unanimously held.

The question is thoroughly known: it is treated in all works on geometry, and for a long time there has been a final judgment upon it. But it is one of those subjects that, by their very subtilty, try some minds; and quibbling is as natural in geometry as in philosophy. This is why the question of "Euclid's postulate" comes periodically before learned societies, as it comes up in the conversation of those who are engaged in teaching mathematics.

In the session of the Academy above mentioned, M. Lionville recalled the fact that many demonstrations of this famous proposition have been attempted, and that none have succeeded, because there are limits beyond which human reason ceases to be authoritative. M. Lionville, indeed, denied that "Euclid's postulate" should be classed among those whose discussion the Academy forbids; that is, the squaring of the circle and the trisection of an angle. He told, in this connection, an anecdote of Lagrange. This great mathematician, thinking he had discovered an exact solution of "Euclid's postulate," went to read his demonstration before the Academy; but, having reflected further on the subject, he changed his mind, and decided that it was best not to do it. He put his manuscript in his pocket, whence it never emerged.

Several geometricians of the Academy, Messrs. Ch. Dupin, Bienaymé, and Chasles, spoke on this occasion, and confirmed the opinions of M. Lionville.

Moreover, when the demonstration of the axiom forwarded by the provincial professor was examined, it was found to be false.

It must be understood and declared, therefore, that axioms in geometry cannot be demonstrated.

There are some good men who have tried to draw from this discussion an argument against the certainty of geometry. One of these is M. Bouilland, the eminent physician, and himself a member of the Institute, who could not, he said, overcome his astonishment at hearing it said that there were several geometries, and that the very foundations of the science were put in doubt. Take courage, great and honorable Doctor: geometry has nothing to lose, nothing to conceal, and the certainty of its methods is not at stake in this question. What is at stake is simply the methodical, standard teaching of geometry. The question discussed was the best mode of informing the mind as to the principles of the science. But as to geometric truths, as to the facts themselves, they are sheltered from all uncertainty. All these quibbles about truths which must be accepted as axioms, or demonstrated like theorems, are merely the vain and sophistical prejudices of disputants. Not a trace of them remains when the facts and mathematical deductions are brought into practice. Ask astronomers who calculate the orbits of the stars, who determine with rigorous precision the moment of eclipses, occultations, &c.; ask those who have calculated the parallaxes, — a very difficult matter, — if they were very anxious to know how it could be demonstrated that the angles of a triangle are equal to two right angles. In practical application, one gets rid of all these subtleties of the schools.

If we may ignore altogether these fault-finding mathematicians who amuse themselves by contesting the axioms of geometry, we may do likewise with some sophists who desire to dispute the axioms of philosophy and reason, and especially the principle of an immortal soul existing in man. Let us give them their say, and proceed on our course. To philosophize beyond measure, to argue mercilessly, is often absurd.

Second Objection. — *We have no recollection of having lived before our entrance on this life.*

This, we admit, is the greatest, most serious argument against our doctrine. But we must immediately add that, if it did not exist, if we had really recollection of a life anterior to this, the doctrine of a plurality of existences would not need the support of such proofs as we demand of reasoning, of observed facts, and of logical induction. It would be obvious, self-evident. Our only merit, our only aim, in this work, is in the attempt to prove the plurality of existences, when we have no recollection of our past lives.

We have already incidentally treated this question. It would not be unprofitable to repeat here all that we have said in other chapters to explain the absence of this recollection.

We said that, if the soul is human at its first incarnation, if it proceeds from a superior animal, it cannot have memory, since that faculty is very feeble in the animal. In the case of a second or third incarnation, the difficulty is serious, for it involves the forgetting of his former life by a man who has lived and is born again.

But, first, this oblivion is not absolute. It has been remarked that there are always in the human soul some effects of impressions received before terrestrial life. Our natural aptitudes, special faculties and vocations, are traces of impressions received long before, of knowledge already acquired, and which, betraying itself from the cradle, can be accounted for only on the hypothesis of a former life. We have lost the memory of facts: but there remains to us its moral consequence, its resultant, its philosophy, so to speak; and in this way are explained the "innate ideas" marked out by Locke, and which exist in our souls from birth, and also the principle of causality which teaches us that every effect has a cause. This principle can be derived only from facts, for an abstraction can base itself on nothing but concrete facts, accomplished events; and this abstraction, this metaphysical idea, that we bring into the world, implies anterior facts. This anteriority can be traced to nothing but a past life.

We have already said that when the soul, relieved from its preoccupations, suffers itself to lapse freely into reverie, we catch glimpses, in the dim distance, of mysterious and vague sights that seem to belong to worlds not wholly strange to us, and which yet resemble in no wise the sights of earth. In this mystical contemplation there is something like a confused recollection of a former life.

The love that we feel for flowers, plants, and vegetation is perhaps, also, we have said, a kind of grateful recollection of what was our first spring.

In conclusion, if these considerations be not accepted as valid, we will say that there is another which, in our judgment, perfectly explains man's failure to remember his former life or lives. It is, we think, according to a definite intention of Nature that the memory of past lives is denied to us on earth. M. André Pezzani, the author of an excellent work, "The Plurality of the Soul's Lives," which has been very serviceable to us in our investigation of this question, replies in these words to the argument of man's lack of memory of a past life: —

"The earthly sojourn is only a new probation, as was said by Dupont de Nemours, that great writer, who, in the eighteenth century, outstripped all modern thought. Now, if this be so, is it not plain that the recollection of former lives would seriously hinder probations, by removing most of their difficulties, and consequently of their deserts, as well as of their spontaneity? We live in a world where free-will is all-powerful, the inviolable law of advancement and progress among men. If past lives were remembered, the soul would know the significance and import of the trials which are reserved for it here below: indolent and careless, it would harden itself against the purposes of Providence, and become paralyzed by the hopelessness of mastering them, or even, if of a better quality and more manly, it would accept and work them out without fail. Well, neither of these suppositions is necessary. The struggle must be free, voluntary, safe from the influences of the past; the field of combat must seem new, so that the athlete may exhibit and practise his virtues upon it. The experience he has already acquired, the forces he has learned how to conquer, serve him in the new strife; but in such a manner that he does not suspect it, for the imperfect soul undergoes re-incarnations in order to develop the qualities that it has already manifested, to free itself from vices and faults, which are in opposition to the ascensional law. What would happen if all men remembered their former lives? The order of the earth would be overthrown: at least, it is not now established on such conditions. Lethe, as well as free-will, is a law of the actual world." *

It will be answered to this that identity is destroyed, if there is no memory. It will be said that expiation, in order to be profitable to a guilty soul, must be accompanied by the recollection of sins committed in a former life; and that he is not punished who does not know why he is punished. But take notice that it is a question less of expiation, as theologians understand it, than of the establishment in a new home, where the soul may take up again the interrupted

* The Plurality of the Soul's Lives. 18mo. Paris, 1865. Third Ed., p. 405.

course of its perfection. In the passage quoted from M. Pezzani, there is too strong a savor of the Catholic dogma of expiation. We are not here to expiate our sins: the word and the idea of sin are old conceptions of Christianity; but they have no foundation in Nature. In our opinion, the earthly life begins here, in order that the improvement of the soul, which has failed to make it in a former ill-spent life, may be resumed and carried out successfully in a new career. But in all this there is neither sin, according to the religious term, nor expiation of sin.

Let us add that the recollection of our former lives, denied to us on Earth, will return when we reach the blest ethereal abodes, where the existences which follow that of Earth will unfold themselves to us. Among the perfections and powers that will belong to the superhuman, will be the memory of all our anterior lives. The superhuman, soaring in the quiet fields of ether, will see painted on his memory, magnified and developed in incomparable strength, all that he did on Earth. He will remember all his actions. Identity will be born again to him. Eclipsed for a moment, his individuality will come back to him, with his conscience and his freedom.

Thus Jean Reynaud, in his admirable book, "Heaven and Earth," paints to us the wonders of that memory restored to man at the end of the many mutations of his being: —

"The entire restoration of our memory seems fairly to be one of the principal conditions of our future happiness. We can fully enjoy life only in becoming, like Janus, kings of time, and knowing how to concentrate in ourselves knowledge of the future and the past as well as of the present. Then, if the perfect life is one day given to us, perfect memory will be granted at the same time. And now, if we can, let us paint the infinite treasures of a mind enriched by the recollections of an innumerable series of lives, each different to the other, and yet wonderfully linked together by continual dependencies. To this marvellous girdle of metempsychoses traversing the Universe, with a jewel in each world, add still, if this prospect seems worthy of our aspirations, a clear perception of the special influence of our life on the subsequent changes in each of the worlds that we shall have successively inhabited: we exalt our lives in immortalizing them, and nobly marry our history with that of heaven; we gather, confidently, because the omnipotent goodness of the Creator has invited us to do it, all the materials necessary to happiness, and build of them the life that the future reserves for virtuous souls; we plunge into the past, by faith, sure of the brightest illumination, as by it we plunge into the future; we banish from Earth the idea of disorder, in opening the portals of time beyond birth, as we banished the idea of injustice when we opened other portals beyond the grave; we stretch ourselves in duration in all directions, and, despite the obscurity that weighs on our two horizons, we lift fearlessly our earthly life above the imperfect existence of those chosen of Christ, who have cast off hope, and whose memory is but a point in the abyss of eternity; we glorify the Creator in glorifying ourselves, God's servants on Earth; and reflect with a holy pride, in contemplating the divine characteristics of our human life, that we are here below, the younger brothers of the angels!" *

In what condition does our soul resume the memory of all its past? Jean Reynaud marks out two periods: first, that which is passed, as the Druids say, in the world of journeyings and trials, of which the Earth is

* Earth and Heaven, last page, Book I. This passage has been revised and abridged by the author, in the edition of his Select Works, published in 1866.

part; second, that in which our soul, freed from its miseries and vicissitudes, follows its destiny in the ever-broadening cycle of happiness, and which extends beyond the Earth. In the first period there is an eclipse of memory, *at each passage into a new medium;* in the second, whatever may be the displacements and transfigurations of the personality, memory is preserved, and endures full and whole.

This theory of Jean Reynaud is admitted by Pezzani, in the work from which we have quoted.

Saving this *eclipse of the memory at each passage into a new medium*, which seems to us hardly comprehensible and useless, we believe with Jean Reynaud and Pezzani that unimpaired memory of our former lives will return to the soul when it reaches the ethereal regions, — the abode of the superhuman. Only in this way is explicable, according to our system, man's lack of memory of anterior existences.

Thus the argument of the want of memory does not stand unanswered. Earlier writers who have thought on this question have already found the solution that we here give. This objection is not such as to shake our faith in the doctrine of the plurality of existences.

We conclude, with M. Pezzani, that it is according to a definite purpose of Nature that man during this life loses the recollection of what he has been. If we did remember this, if we had before our eyes (as in a mirror) all that we had done in our former lives, our career would be sadly disturbed by the recollection, which would trammel most of our actions, and deprive us utterly of free-will.

Why do we fear death? Whence this invincible dread of the last hour, which comes to all men? Death is after all not very terrible, since it is not the end, but a mere change of state. Man feels such a painful dread of death, because Nature imposes this feeling upon him, with a view to the conservation of his species. So Nature suppresses in man the recollection of past lives, to leave him free to pursue, untrammelled, his new career, and not to cramp him in his actions while he undergoes the ordeal to which he is subjected on this Earth.

Hence the fear of death and the non-recollection of our former life are, in our opinion, referable to the same cause. The first is a salutary illusion, which God puts upon the feebleness of humanity: the second is intended to secure the freedom of its actions.

Third Objection. — Another objection will be made to our theory. It will be said: *That the reincarnation of souls is not a new idea: it is, on the contrary, as old as humanity. It is the metempsychosis, which passed from the Indians to the Egyptians, from the Egyptians to the Greeks, and was afterwards believed by the Druids.*

It is true that metempsychosis is the oldest of philosophic conceptions: it is the first theory devised by man to explain the first cause and the destiny of mankind. We do not see in this remark an objection against our system of Nature, but rather a confirmation of it. An idea does not traverse the ages, accepted and professed through five or six centuries by the elect of different generations, unless it rests on a solid basis. There is nothing, then, against our finding our opinions to be in agreement with philosophical ideas that date back to the remotest times of the history of peoples. The first observers, and especially the Oriental philosophers, — the most ancient thinkers whose writings have come down to us, — had not, like us, minds weakened, prepossessed, and misled by routine, trammelled by the *dicta* of masters. Placed close to Nature, they seized her real aspects, taking nothing from education or schools. We must, therefore, congratulate ourselves on discovering that we have reached, by logical deduction, the ancient conception of Indian wisdom.

We ought, however, to note a material difference between our system of the plurality of lives and the Oriental dogma of metempsychosis. By the Indian philosophers as by the Egyptian, and in the Greek school, which inherited the maxims of Pythagoras, it was admitted that the soul leaving the human body passed into that of an animal, by way of punishment. We reject utterly this useless retrogression. Our metempsychosis is ascending and progressive: it never descends, never goes backward.

A rapid view of the doctrine of animal metempsychosis, as it was professed in the different philosophic sects of antiquity, will not be out of place here. It will help us to understand how it differs from our system, and to see at the same time how popular metempsychosis was in ancient times, in Europe as well as in Asia.

The most ancient book known is probably the Vedas, which embodies the religious principles of the Indians or Hindoos. In this compendium of the first religions of Asia is found the general doctrine of the final absorption of souls in God. But before reaching this mergence in the Great All, the human soul has to pass through life in all its activity. It made therefore a series of transmigrations and journeys, in different places, in different worlds, and through the bodies of many different animals. Men who have not done good works go to the Moon, or the Sun, or even return to the Earth, to enter the bodies of certain animals, like dogs, butterflies, worms, and adders. There are also intermediary places, between the Earth and the Sun, where souls that have not very seriously offended go, to pass a season of probation. The Purgatory of the Catholics was borrowed from the religion of the Hindoos.

Let us quote, in support of this general statement, some passages from the Vedas: —

> "If a man has done deeds that lead the way to the world of the Sun, his soul goes to that world: if he has done deeds that lead to the world of the Creator, the soul goes to the world of the Creator." *

The book of Vedas declares very distinctly that the animal, as well as man, has the privilege of passing to other worlds, in reward for his good works. Oriental wisdom, it seems, did not entertain the unmerited

---

* The Religion of the Hindoos, according to the Vedas. By Lanjuinais.

contempt for animals which modern religion and philosophy bestow on them.

"All animals, according to the degree of intelligence they had in this world, go thence to other worlds. . . . The man who looked to the reward of his good works, when he dies goes to the world of the Moon. There he is at the service of the officers of half the Moon, in its crescent. They receive him joyfully. There is no rest, no happiness for him: all his reward consists in his having come for a time to the world of the Moon. This time past, the servant of the officers of the Moon in its crescent descends to hell. There he is born again a worm, a butterfly, a lion, a fish, a dog, or in some natural form (even under the human form).

"In the last stages of his descent, if he be asked, 'Who are you?' he answers: 'I come from the world of the Moon, the price of works done in the hope of reward. Here I am clothed anew with a body. I suffered in the belly of my mother, and since I came from it. I hope at last to acquire that knowledge which is every thing; and to enter into the right way of worship and meditation, without thought of reward.'

"The world of the Moon is where is bestowed the recompense of good works done without renunciation of their fruits, their merits; but this recompense is of fixed duration, after which the soul is born again in an inferior world, a bad world, a world for the punishment of evil.

"On the other hand, by renouncing all pleasure and the reward of works, and seeking God with steady faith, the soul comes to the Sun, which is endless, which is the great world, and whence it returns not into the world for the punishment of evil." *

The Egyptians, who borrowed this doctrine from the Hindoos, made it the basis of their religious system. Herodotus tells us † that, according to the Egyptians,

---

* The Religion of the Hindoos, according to the Vedas, pp. 324, 325.

† Histories, Book II. chap. xxiii.

the human soul, passing from a completely decomposed human body, enters the body of some animal. It occupies three thousand years in passing from the body of an animal through a series of others: at the end of this time the same soul, returning to human kind, enters the body of a new-born child.

The Egyptians took especial care in the conservation of human bodies. They embalmed the bodies of their relatives or state officials, and thus prepared the mummies that every one has seen in our museums. The object of this universal custom of embalming was not, as one would naturally think, to hold the body in readiness to receive at the end of three thousand years the soul that was to be incarnated in its original residence. In fact, the Egyptians believed that the soul would take up its residence not in the old body, but, as we just said, in that of a new-born child. Embalming, then, had another object. It was supposed that the soul abandoned the human body, to begin its migrations through the bodies of animals, only after the corpse was entirely decomposed. Hence the efforts of the Egyptians to delay the moment of this separation, by preserving the corpses from destruction as long as possible. This is what Lewins tells us: —

"The Egyptians, renowned for their wisdom, kept the corpses of the dead, in order that the existence of the soul, linked with that of the body, might be preserved, and not escape as speedily as in other countries. On the other hand, the Romans burned corpses, in order that the soul, regaining its liberty, should re-enter instantly into Nature." *

* Commentaries on Virgil, lib. iii.

The oldest and most extraordinary of the Greek philosophers, — we mean Pythagoras, — on his travels in Egypt, picked up the idea of metempsychosis. He brought it into his school; and all Greek philosophy, which formed itself on the lessons of the sage of Crotona, believed, with him, that the souls of the wicked passed into the bodies of animals. Hence the abstinence from flesh which Pythagoras enjoined upon his disciples, — a notion which also he had picked up in Egypt, where the respect with which animals are regarded is due to the general belief that the bodies of beasts were inhabited by human souls, and that consequently in abusing an animal one ran the risk of maltreating one of his ancestors.

The philosopher Empedocles adopted the system of Pythagoras. He says, in his lines quoted by Clement of Alexandria, —

> "And I, also, I was a young girl,
> A tree, a bird, a dumb fish in the depths of the sea."

Plato, the most illustrious of the Greek philosophers, in his sublime ideas about the soul and immortality, gave a prominent place to the system of Pythagoras. He granted that the human soul passed through certain animals, by way of expiation of its crimes. Plato said that we remember on Earth what we did in our former lives, and that to learn is to remember.

We read in the Timæus, —

"Cowards are changed into women; trifling and vain men into birds; the ignorant into savage beasts, grovelling and bent to the ground according as their sluggishness while on Earth was more or less degraded; souls stained and corrupt go to animate fishes and aquatic reptiles."

In the Phædon we read, —

"Those who have abandoned themselves to intemperance, to lasciviousness or gluttony, having no continence, enter, it is likely, into the bodies of animals like themselves; and those who loved nothing but injustice, tyranny, and plunder, go into the bodies of wolves, hawks, and falcons. The destiny of other souls is according to the lives they have led." *

Plato reduced the time allowed by the Egyptians for the passage of the soul through the bodies of animals. Instead of three thousand years, he fixed it at only a thousand. Yet he would have this thousand years repeated ten times, which would give a total of ten thousand years for accomplishing the entire circle of existences. Between every two of these periods, the soul made a short stay in hell. During this nether residence it drank the water of the Lethe, in order to destroy the memory of its former life and begin the new life without any recollection of its predecessors.

Plato exalted the dogma of animal metempsychosis by grand views of spiritual immortality and the freedom of man, — views that are quoted even in our day with fervent admiration; but it would take too long to reproduce them here.

Metempsychosis held a place in Plato's philosophy inferior to that it had occupied in Pythagoras's and

* Phædo, Plato's Works, translated by M. Cousin, Book I. p. 242.

in the religion of ancient Egypt. It regained all its importance in the philosophies of the school of Alexandria, which perpetuated in Egypt the school and traditions of the Platonic philosophy, and revived in the land of the Pharaohs the days of the Lyceum at Athens.

Plotinus has given us in his Enneads a long amplification of Plato's doctrines. Here is what this commentator says of the doctrine of the transmigration of souls: —

> "It is an opinion recognized from the earliest time that, if a soul sins, it is condemned to expiate its sins *by suffering punishment in Tartarean darkness; then it is permitted to enter a new body to begin anew its trial.*"

This passage proves that in the esteem of the ancients this sojourn in hell was merely temporary, and that it was always followed by new trials, more terrible and painful according to the nature of the sins to be atoned for. Plotinus also says, —

> "When we are led away in the multiplicity [which means that when we are wedded to matter and bodily passions], we are punished first by our misconduct itself: when we resume the body, we have a less happy condition."

In another passage, Plotinus thus expresses himself about the transmigration of souls: —

> "The soul leaving the body becomes that power which it has most developed. Let us fly, then, from here below, and rise to the intellectual world, that we may not fall into a purely sensible life, by allowing ourselves to follow sensible images; or into a vegetative life, by abandoning ourselves to the pleasures of physi-

cal love and gluttony: let us rise, I say, to the intellectual world, to intelligence, to God himself.

"Those who have exercised human faculties are born again men. Those who have used only their senses go into the bodies of brutes, and especially into those of ferocious beasts, if they have yielded to bursts of anger; so that, even in this case, the difference between the bodies that they animate conforms to the difference of their propensities. Those who have sought only to gratify their lust and appetites pass into the bodies of lascivious and gluttonous animals. Finally, those who, instead of yielding to their lust or their anger, have rather degraded their senses by disuse, are compelled to vegetate in the plants; for in their former life they have exercised only their vegetative power, and have striven only to become trees. Those who have loved music to excess, and have yet lived pure lives, go into the bodies of melodious birds. Those who have ruled tyrannically become eagles, if they are otherwise free from vice. Those who have spoken lightly of heavenly things, keeping their eyes always turned toward heaven, are changed into birds which always fly toward the upper air. He who has acquired civic virtues becomes a man: if he has not these virtues in a sufficient degree, he is transformed into a domestic animal, like the bee, or any other of like kind." *

Every one knows that among our ancestors, among the Druids and the high-priests of the Gauls, metempsychosis was professed, almost as the Egyptians and Greeks had understood it. There is therefore something national in it to us. It has been honored, and its principles have flourished, in the very countries where we live.

We have recalled these facts, and quoted from ancient writers, merely in order to state precisely the understanding that the Egyptians as well as the Greeks,

---

* Enneads of Plotinus, translated by Bouillet.

and later the Gaulic priests, had of metempsychosis. Our system differs from the old Oriental conception, which was adopted by the Egyptians and Greeks, and afterwards by the Druids, in this, that we do not admit that the human soul can ever return into the body of an animal. It has, we believe, passed through such a preparatory medium, but it returns not to it. The animal, in fact, has but an inferior *rôle* in Nature,— inferior to that of man: it is below our kind in its degree of intelligence, and can have neither merit nor demerit. Its faculties do not give it entire responsibility for its acts. It is only an intermediary link between the plant and the man: it has some faculties, but it cannot be pretended that they assimilate it to moral man.

Therefore we reject this return of the human soul by ways which it has already traversed. Retrogression is not in our doctrine. The soul may pause an instant in its onward march, but it never goes back. We grant that man is doomed to begin anew an ill-spent life; but this new trial takes place in a human body, in a new envelope of the same living type, and not in the body of an inferior being. The Oriental dogma of metempsychosis ignores the great law of progress, which is, on the contrary, the foundation of our doctrine.

Fourth Objection.—Yet it will be said: *You contend that our soul has pre-existed in the body of an animal: do you share the opinion of naturalists who derive man from the ape?*

Certainly not. French and German naturalists who,

applying to man the theory of Darwin about the transformation of species, have derived man from an ape, appeal to anatomical arguments exclusively. Messrs. Vogt, Buchner, Huxley, and Broca compare the skeleton of the ape with the skeleton of the primitive man; they study the form of the cranium of both; they examine the depth of the lines which serve for muscular insertions in the thigh-bone (that is, the *ligne apre* of the femur); they measure the size and prominence of jaw-bones, &c. From these comparisons they draw the conclusion that man is derived anatomically from a species of quadrumanous animal. The soul is not taken into account by these *savants*, who reason just as if there were nothing that thinks in the anatomical cavities they explore and measure. On the contrary, we reach our conclusion by comparing the faculties of the human soul with those of animals. To us animal forms are nothing: the spirit, in its different manifestations, is our principal objective.

Why, indeed, should we derive man from an ape, rather than from any other mammiferous animal, as the wolf or the fox? Does any one think there is much difference between the skeleton of an ape and that of a wolf, of a fox, or any other carnivorous animal? Put together these three or four skeletons, and it will not be easy to distinguish one from the other, if, instead of an ape (which belongs to the higher species), you take a quadrumane of an inferior species, — a lemur, a striated monkey, or a dog-faced baboon. Compare the physiological functions of a fox and a wolf with those of an ape, and you will find just alike, in all these animals, the functions of digestion, respiration, the circulation of the blood, the lymph, the nervous organization, &c. Examine the organs that serve these functions, and you will see that they are identical in structure in all these animals. Why, then, do you make man come from the ape rather than from the wolf or the fox? Is it because the apes in our menageries wear a distant resemblance to man in their somewhat vertical posture, and in certain features of their physiognomy which caricature ours? But, in the great family of apes of the two worlds, how many kinds present these characteristics? Scarcely five or six. All the rest of the ape family have the bestial muzzle fully developed, and prove themselves inferior in intelligence to most other mamniferous animals. If you derive, from an organic point of view, man from the ape, because certain species of these quadrumanes caricature man in their faces, why not derive him from the parrot, since this bird utters articulate sounds which caricature the human voice? or even from the nightingale, because this melodious songster of the woods modulates sounds like our cantatrices?

The consideration of animal forms has very little interest for us in determining the place that a living being is going to occupy in the scale of creation: because forms have the same type in all superior animals; because the body varies very little in structure throughout the great class of mammifers; because physiologic functions proceed in the same way in all.

So the basis that we take for our researches is quite other. It is a spiritual basis: in the faculties of the soul we seek means of comparison.

It cannot be urged that we espouse the doctrines of Darwin and the transformists, because we contend that the soul has been domiciled in the bodies of several animals before reaching a human body; because we admit that the spiritual principle begins in plants in the germ state, and that this germ develops and grows while passing through the progressive series of animal species to end in man, where it is to finish its elaboration and improvement. The Darwinists, or transformists, regard only anatomical structure, and make an abstraction of the soul. We are guided not by the materialistic idea, which directs and inspires these *savants*, but, on the contrary, by a rational spiritualism.

Fifth Objection. — *This system is borrowed from Fontenelle's work*, "*The Plurality of Worlds.*"

Our point of departure was, indeed, the work of Fontenelle; but we have made large additions to the ideas contained in that justly famous book. It will be well, however, to state here what development we have given to the thought of the immortal author of "The Plurality of Worlds."

The discoveries of Newton, completing the mathematical labors of Kepler, and establishing firmly the system of modern astronomy, had scarcely come to an end, and had scarcely become known to the scientific world, when it occurred to Fontenelle to popularize, as the modern saying is, the works of Newton, — to communicate to all enlightened minds a knowledge of the new system of the world, a triumph which was instantly seen to be of fundamental importance to the sciences and philosophy. An excellent writer, of a sharp and subtle mind, a poet and man of letters, a worthy nephew of the two Corneilles, Fontenelle possessed a literary arsenal marvellously well furnished. He brought all his intellectual resources to a work which appealed at once to the world at large and to *savants;* and composed that masterpiece of intelligence, acuteness, and grace, entitled "The Plurality of Worlds," which appeared in 1686.

In a series of conversations between a conventional marquis and an imaginary chevalier, Fontenelle explains, with admirable clearness, and sometimes with the touches of a finished comedian, the true arrangement of the worlds. He expounds the system of Copernicus, Kepler, and Newton; that is, the fixity of the Sun and the stars, the movements of the planets around the Sun, the rotation of the Moon around the Earth, the phases of our satellite, &c. Then, entering upon another subject, he tries to prove that the Moon and the planets must be inhabited. He makes it plain that all the planets that move around the Sun are like the Earth; and that they must be, like it, theatres of life, — that they must furnish homes to animals and men. He adds, however, that the planets being warmer or colder than the Earth, according to their distance from the Sun, and according to the presence or absence of an atmosphere around them, matter must take on other aspects in these media, determined by external influences. All of which means that the living beings who inhabit the planets must be constituted very differently to the men and animals which people the Earth.

"The Plurality of Worlds" has been one of the most admired works in our literature, and it well deserves such appreciation. It would be impossible to lend a greater grace to science, to be more attractive, more amusing, and more ingenious in treating of physics and astronomy.

Fontenelle's book has been for two centuries constantly copied, turned inside out, and repaired by many writers, who have adapted the author's ideas to the literary form of their times, but who have added almost nothing to the substance of the matter.

In Fontenelle's day, an illustrious Dutch *savant*, resident in Paris — Christian Huygens — the inventor of clocks moved by weights and springs, did not disdain to make over "The Plurality of Worlds," which had first seen the light about a dozen years before; and in 1698 appeared Huygens's "Cosmotheoros,"* a work which, excepting the form, reproduced bodily the ideas of the spiritual conversations of Fontenelle.

Like Fontenelle, Huygens described the new astronomical system according to the investigations of Newton; then he undertook to prove that the planets are

---

* *Κοσμοθεωτός*, sive de Terris Cœlestibus earumque Ornatu Conjecturæ. A posthumous work.

inhabited, as well as the Earth. The treatise of the Dutch *savant* is more profound, scientifically regarded, than that of the amiable talker of the Conversations, and contains some very remarkable scientific considerations as to the habitability of the planets, which more than one modern author has made good use of.

In our century, works modelled on Fontenelle's "Plurality of Worlds" have been very numerous in France and other countries. The list of these works, mere imitations of the fundamental work of the French writer, would be too long to have place here.

The last imitation of Fontenelle's book, that has made any noise, was the work of an English naturalist, David Brewster. In 1833, a *savant* who held an ecclesiastical post in England, Mr. Whewell, had published a dissertation directly contradicting the theory of Fontenelle. In a work entitled "An Essay on the Plurality of Worlds," * he endeavored to prove that the doctrine of the plurality of worlds was contrary to religious faith and to science. The Bible is, it is well known, the grand oracle, the religious fetish of the English. Now the Bible says nothing of beings who live in the planets. It does not even speak of planets, for the reason that in its day planets were not distinguished from the mass of stars with which these globes are confounded, owing to their aspect in the vault of the

---

* M. Figuier translates this title thus: "On the Plurality of Worlds an Essay," regardless of the potentialities of punctuation. — Tr.

heavens. Consequently the theory of the plurality of worlds was an offence against Christian faith, said Mr. Whewell, who, on the same occasion, sought to prove that Science not less sternly condemned it.

The English philosopher, David Brewster, took up the defence of astronomy. He published an admirable *brochure*, in which he combated the sayings of the reverend gentleman, by arguments borrowed from Fontenelle and Huygens, and by new ones furnished by the recent advances of contemporary science. David Brewster, in the work which is entitled "More Worlds than One; the Philosopher's Faith and the Christian's Hope," proved that this doctrine is as religious, as Christian, as it is scientific. This controversy throughout deeply interested our friends across the channel.

A French writer, M. Flammarion, then had the notion of publishing a work on the habitability of the planets, in which he referred to the polemic storm stirred up in England about this question, by the writings of Brewster and his opponent. This work was called "The Plurality of Inhabited Worlds." *

So it is Fontenelle to whom must be reserved the very great merit and honor of having imagined the theory of the habitability of the planets, and rendered probable the existence of man and animals in those distant worlds. Fontenelle thinks that the planets contain living beings like man. He sees no reason, since our Earth is inhabited, and differs not from the

---

* 1 vol. 18mo, Paris, 1865.

other planets, why these stars should not also give a home to organic life, to motion, to feeling, and to thought.

Yet neither Fontenelle nor Huygens, nor their innumerable imitators, have ever taken the trouble to inquire whence came, or whither go, the inhabitants of the planets. They do not connect their existence in the planets with any view of the whole. They do not ask what relations subsist between planetary beings and those of our globe. It is on these points that we have made, in this work, large additions to the ideas of Fontenelle and Huygens, and their imitators.

Several writers who have exercised their imagination on this subject have said that souls journey from planet to planet, pausing for a sojourn on each, and seeing at every station their moral improvements and the measure of their happiness increase. Such was the opinion of Charles Bonnet, Dupont de Nemours, and Jean Reynaud, shared in our day by a certain number of thinkers. The astronomer Bode has written that we start from the coldest planet of our solar system (Uranus), and advance progressively from planet to planet, ever drawing near the Sun. In the Sun will live, in the opinion of this astronomer, the most perfect beings in creation. But, in direct opposition to this theory, the German philosopher, Emmanuel Kant, in his "General History of Nature," says that souls start from the Sun in a state of imperfection, and travel from planet to planet, going farther and farther from the Sun. To Kant, Paradise would be not in the Sun, as Bode would have it, but, on the contrary, in a planet which is the coldest and the most distant from the central star of our system. Between these two extremes are arrayed all the theories that *savants* or imaginative writers have formulated, more or less clearly, as to the journeys of souls through the planets.*

But we pay no attention whatever to these views, because we believe that souls find in all the planets the same conditions of habitability; and because, in our opinion, Nature is nearly the same in all the planets, and that there are in all, as in our globe, vegetables, animals, and a being superior in intelligence and moral power to all the rest of the living creation, the being whom we call planetary man. Whatever their distance from the Sun, the planets are inhabited, we believe, by animal series, like those that live on our globe. At the death of the terrestrial, or of the planetary man, the human soul passes into the ether which surrounds all the planets, and which reaches even to the Sun.

No one before ourselves has thought of regarding the ether as a medium that reborn human souls could live in, and no one has hit upon the idea of making the ether in which move the stars of our system the general rendezvous, the common home, of beings superior to planetary humanity. No one has declared that in this ethereal space, which is coextensive with our solar sys-

---

* The list and analysis of all these works, sufficiently tiresome withal, may be found in a book by M. Flammarion, "Les Mondes Imaginaires," Paris, 1869.

tem, living beings may undergo successive transformations and repeated incarnations, up to the moment when the soul, purified of all earthly alloy, reaches the Sun, which we hold to be the ultimate abode of beings wholly spiritualized, absolutely immaterial.

Our opinions may find no sharers; our system of Nature may be unfavorably criticised or rejected. We present it as a purely personal view, and would not impose it on any reader. The merit, if there be any, of this philosophic and scientific conception, consists in the vast synthesis in which we bind together all the living creatures that people the solar world, from the smallest plant in which appears the form of organization to the animals, from the animal to man, and from man to that series of superhuman and arch-human beings who dwell in the ethereal spheres, and finally from these latter up to the radiant inmates of the solar star. The question of the habitability of the planets has its place in our system; and Fontenelle's book, "The Plurality of Worlds," has furnished us some very useful facts for this part of our work. But this place is not the first. There are scientific considerations quite as important, which have aided us quite as much as Fontenelle's ideas, in building up our doctrine of the plurality of existences and reincarnations. It is by combining, on the one hand, all the discoveries of modern chemistry as to the composition of plants and the physical phenomena of their respiration, — and, on the other, what is certainly known as to the physical and chemical properties of solar light, — that we have conceived the idea of making sunlight the vehicle of animated germs which are deposited in plants by its rays. By pondering upon what the philosophers Charles Bonnet, Dupont de Nemours, and Jean Reynaud, have written as to the physical conditions of re-born human beings; by invoking our own meditations on the destiny of man beyond the formidable barrier of the tomb; in a word, by appealing to the most diverse sources that science and philosophy can offer to us, we have composed this essay of a new philosophy of the Universe.

This system, we repeat, may be erroneous, and another more logical or wiser may be substituted for it. But what we hope will remain is the synthesis that we have realized of all the facts of the physical and moral order here brought together. It is the bond by which we unite one to another all the beings of creation, and which comprises both the organic and moral attributes of these beings; it is this great ladder of Nature, on the rounds of which we place every thing that lives; this circle without an end, by which we weld together all the links in the chain of living beings. The theoretic explanation that we have formulated of all the facts thus grouped together may not be accepted, we repeat; but we believe that the facts are well collected, and that upon their grouping must be based any theory that undertakes to explain the Universe. If our exposition is contested, our synthesis of the facts will stand, we hope.

And, moreover, it is only in this way that the sciences

have been pushed forward,—the exact as well as the moral sciences. Chemistry was not, as has been asserted, created by Lavoisier: it was founded by Stahl. It was not the pneumatic theory put forth by Lavoisier, but rather the system of phlogistics devised by Stahl, that instituted chemistry in the last century. Stahl, it is known, had the great merit of uniting all the facts known in his day in a general theoretic exposition, of oomposing a whole with them, and of creating the system of phlogistics. This system was inexact, no doubt; but the facts that he failed to combine in building it had been perfectly well chosen, and excluded no useful element of information or research. So when Lavoisier came, forty years after Stahl, he had only to turn the system of his predecessor inside out, as one turns a coat. For phlogistics, Lavoisier substituted oxygen. He preserved all the facts, only changing their exposition, and chemistry was founded.

A well-constructed synthesis must necessarily precede every theory of Nature. Descartes, in elaborating his system of vortices, formulated a very inexact conception; but the facts on which this theory rested were so well chosen, and answered so precisely to the needs of science, that when Newton came forward with his system of attraction he had only to apply the new hypothesis to the facts gathered by Descartes for his vortices, in order to produce the true astronomy and the true science of Nature. When Linnæus created his system of botany, he made a division of vegetables, which was certainly very artificial; and he himself recognized all its defects. But, thanks to this artificial method, he succeeded in grouping all plants in a methodical catalogue. If the principle of classification was vicious, the service which the same catalogue rendered to botany was very great. In fact, only by starting from Linnæus could one tell where he was in the confused maze of facts that one had to gather and fix in his memory, in order to be able to prosecute the study of vegetables. Botany really took its first flight from the publication of "The System of Nature," by the immortal botanist of Upsal.

Our pretension in this work is not to propound an unassailable theory of the Universe, but merely to combine and group methodically the facts on which such a theory must rest, — facts physical as well as metaphysical and moral.

CHAPTER XXII.

Continuation of Objections. Some Persons cannot understand how the Rays of the Sun, Material Substances, can be the Germs of Souls which are Immaterial Substances.

A FIFTH objection will be advanced against our system of Nature. It will be said: *How can solar rays, which are material, transport animated germs which are immaterial? These two terms are incompatible.*

There is in the Holy Scriptures a magnificent simile which we are going to employ, in order to answer the objection or the question that has just been put.

St. Matthew speaks of a grain of mustard-seed, — that is, of the seed of a tree, — which, dropped into the earth, produces an herbaceous plant, and then a tree with spreading branches; and he is astonished to see this insignificant seed produce that imposing inhabitant of our forests, which, covered with flowers and fruit, displays its beauty in the bosom of creation, and offers in its shade a resting-place for weary birds. Not only, says the evangelist, is there nothing in this huge tree which recalls the humble seed from which it sprung, but there is not in the tree a single atom of the matter which originally composed the seed.

This grain of mustard-seed to us is the image of the Sun's rays, which falling upon the Earth sow therein animated germs, which produce plants, which soon will give birth to animals, and afterwards to man, as well as to the whole series of creatures invisible to us that succeed him in the heavenly realms.

It is nothing, apparently, this seed of a tree, — this little cold seed without odor or color. Nothing distinguishes it in appearance from the straw that lies near by. Yet it contains the mysterious leaven, that sacred essence, so to speak, which is called a germ; and what marvels are going to spring from that sacred essence!

In the first medium into which it is cast, — that is, into the obscurity of wet and cold ground, — this germ is transformed: it becomes a new body, without any resemblance whatever to the seed which enfolded it. It yields a plantlet, a being subterranean but perfectly organized, which has its root which burrows in the soil, and its stem which takes an opposite direction. Between these two is the seed, eventerated, burst, having suffered the grain to escape, whose office expires at this point.

The subterranean plantlet is a wholly new being. It has nothing in common with the seed from which it proceeded. It is ternal and colorless, but it breathes: it has canals in which liquids and gases already flow.

Soon the plantlet comes out of the earth. It salutes the light, it appears to our sight, and is then a very different creature from the subterranean individual. The new-born vegetable is not, as it was, beneath the surface of the earth, ternal and gray: it is green. It respires like other vegetables; that is, by yielding oxygen under the influence of the light, whereas it evolved carbonic acid under ground. Instead of the dull and sorry subterranean plantlet, you have a green and tender sprout furnished with peculiar organs. Where is the grain of mustard-seed?

Speedily our sprout grows and becomes a young plant. Still feeble, and hidden under the shade of grass, it has nevertheless a complete individuality. It resembles neither the sprout nor the plantlet, its subterranean ancestors. The sprout shoots up and becomes copse; that is, the stripling of the vegetable kingdom, the fire and mettle of herbaceous youth.

At this point the plant has already changed its entire substance several times, and there remains nothing of the organic and mineral elements that existed in the several beings that preceded it in the same little spot of earth where the changing phases of its strange metamorphoses have passed.

Wait a few years, and you will see the main stem of the copse, having been properly disembarrassed of neighboring sprouts, and restored to its own individuality, stretch up and increase. Its respiration has become quite active. Its leaves wide-spread vigorously suck in the carbonic acid gas of the air. The exhalation of vapor throughout its foliaceous surface goes on with energy. It is a young and stout tree, growing every day stronger and more beautiful.

During this growth, during the transformation of the shrub into the young tree with a single slender stem, a new being has been formed. The organs that it lacked have come to it, and made it a singular individual. It has flowers, bracts, new vessels for circulating the sap and juices which it has not yet elaborated. The surface of its leaves has changed in structure, so that absorption is more powerful.

Where is the sprout, the being whence sprung our young and vigorous tree? What physical relation or resemblance is there between these two beings? We can see nothing but differences. One individual has succeeded another individual. The vegetable has changed, not only in the matter which has changed in it, but in the form of its organs: new forms have followed each other serially in the shrub, since it was a mere sprout living on the surface of the soil.

It is quite another thing when the young tree is full grown; when, in the lapse of years, its trunk has become hardened and incrusted with thick accumulated layers of bark; when its branches have multiplied infinitely; when efflorescence and fructification have thoroughly changed all its parts, internal and external. Then it is the imposing cedar, covering with its majestic and kindly shade a wide extent of ground, the proud oak which spreads afar its robust and knotty branches, or the supple chestnut which pushes in all direction its shining and polished arms. The organs belonging to these luxuriant vegetables, the pride of the forests, have no longer any relation with those which belonged to them in the first years of their life. The flowers that in spring-time crown with white plumes the tops of the branches, the fruits that follow the flowers, the seeds enfolded in the protecting rind of the fruit, — these are peculiarities that make these proud and mighty trees beings unique in Nature.

Where, as St. Matthew asked, is the grain of mustard-seed that once obscurely sucked up the juices of the earth? All is changed, — dwelling-place (for the medium is the air, the earth no longer), form, and physiological functions. And not only is all changed, but it has changed very many times. Not only is nothing left of the matter which composed the shrub in the first part of its life, but nothing is left of the organic forms that belonged to the vegetable in its infancy.

Yet — O mystery! O Nature! — amid all these changes, notwithstanding this constant succession of beings which mutually replace each other, there is something that remains unchangeable, which has never changed, but has preserved its individuality always. This is the secret power that produced all these changes, that directed all these organic mutations. This power is, we believe, the animated germ that the young plant received from the seed whence it came. In all the transformations that the vegetable being has undergone, despite the numerous phases it has passed through, and which have produced a series of different beings, succeeding each other in its material substance, — the spiritual principle, the cause and prime agent of all this prolonged activity, has remained the same. The animated germ which dwells to-day in the full-grown vegetable is the same that was there during its growth, the same that was there in its budding state, the same that slept in the seed that was once shut up in the bosom of the damp cold earth.

In this majestic tree, which, springing from an imperceptible and inferior seed, has seen a whole genealogy of beings succeed and replace each other, different in form and size, — and which, despite its continual transformations and its incessant development, has always preserved the unique and unchangeable principle of its activity, — we see a faithful image of the soul, enduring, unique and indestructible, among the beings or different bodies that it has successively inhabited. Sprung from a germ, it has never ceased to grow, to develop and spread, remaining always itself. The grain of mustard-seed, or the seed of the tree, to our eyes is the plant or the inferior animal in which the Sun has deposited the animated germ. The subterranean plantlet is the animal charged with perfecting the germ transmitted by the plant, and which develops and amplifies it. It is, for example, the fish or the reptile perfecting the spiritual principle received from the zoöphyte or the mollusk. The sprout that coming out of the earth grows in the shade of the grass, and tries its aerial organs, is an animal a little higher in the organic scale, such as a bird; in which the animating principle, proceeding from the reptile or the fish, increases in intellectual power. The young vegetable in the copse state, which lives a purely aerial life, is the mammifer. The tree with the slender trunk, pushing out its young branches, is man improving the soul that he has received from a mammifer. At last the mighty dean of the forest, surpassing all neighboring trees in size and majesty, with its vast top and wide-spreading branches and splendid flowers, is the superhuman who dwells in the bosom of the ethereal fluid; and who, later, will give place to a series of still superior creatures, who will mount from station to station, from one heavenly halting-place to another, even to the shining realm, — that is, the Sun, — where sit enthroned those purely spiritual beings, whose essence is absolute and perfect immateriality.

Thus the animating principle remains unchangeable and identical with itself during all the transformations undergone by beings charged to receive successively this precious deposit, — from the vegetable in which it first made its home in the germ-state, through the series of living creatures, from the plant and the zoöphyte up to man and the superhuman. Despite all these external changes, the same spiritual principle survives unaltered, ever improving and amplifying itself.

Let us complete the similitude. When the forest tree has matured its fruits, they half open, their seeds fall out on the ground, or are scattered by the winds. Falling on damp soil, they germinate; and, according to the laws of Nature, young vegetables spring up, as we have shown. From a single oak, cedar, or chestnut, come a multitude of like trees. Now just as the full-grown tree drops on the ground from its thousand branches seeds that will germinate, so the spiritualized beings who dwell in the Sun shoot upon all the planets their emanations; that is, their animated germs. These are the germs sent upon Earth by the rays of the Sun, and, falling on our globe, produce vegetables, which later will give birth to different animals that we know, by the successive transmigration of the same soul through the bodies of these beings.

We can now answer the objection stated in the caption of this chapter, in this wise: *How can solar rays, which are material, be the vehicles of animated germs, which are immaterial?*

When naturalists accept the Newtonian theory of light, — that is, the theory of emission, — they must necessarily regard light, and consequently the solar rays that produce it, as material bodies. But this theory is now erased by science, and is replaced by the theory of undulations, devised by Malus, Fresnel, Ampère, and the pleiad of great naturalists and mathematicians who made illustrious the beginning of our century. It can no longer be believed that solar rays are, as the partisans of the emission theory would have it, a material emanation from the Sun's substance. Facts gathered from all sources prove that the solar rays are not matter transported from the Sun to the Earth, but that light, like heat, results from the primal shock inflicted by the Sun on the ether which is spread through all space. This shock, a disturbance passing from molecule to molecule, from planetary ether even to our globe, produces the phenomena of light and heat. We need not further develop this idea at this point, in order to explain more scientifically the theory of undulations, which will be found sufficiently demonstrated and elucidated in works on physics. We have only to prove that, according to the principles of modern science, solar rays are not material bodies, but result from a simple disturbance or vibration of the planetary ether. If, therefore, the Sun's rays are not material, there can be no difficulty in admitting that these rays, being immaterial, are the vehicles, the carriers of the animated germs, which also are immaterial.

But if the question be urged still farther, if we be asked to explain more precisely how immaterial germs can travel through space, we answer that one must preserve one's self from the folly of wanting to explain every thing. Absolute explication is forbidden to the feeble reach of our intelligence. We are forced to confess our impotence when it comes to explaining rigidly the simplest phenomena of Nature. What is the true cause of the fall of bodies, of the gravitation of the stars, of electricity, of heat? What causes the circulation of our blood and the beating of our hearts? The deepest obscurity veils the first causes of these phenomena, which nevertheless we daily witness; and the more we long to penetrate their inmost essence, the thicker darkness grows about our minds. So naturalists have laid down, since Newton, a wise and admirable principle. They have agreed to study carefully the laws of physical phenomena, to measure exactly the effects of heat, of gravity, of electricity, or of light, and also never to trouble themselves about searching out the causes of these phenomena. The better informed one becomes, the greater advances he makes in knowledge of the Universe, the more forcibly is he impressed with this truth, — that man knows nothing absolutely as to first causes; that he may think himself happy in knowing the laws by which the effects of first causes manifest themselves, — that is, the physical and vital actions that are plain to our eyes; but that he must impose upon himself, for his own peace, the rule never to seek to find out the *wherefore*, the *pourquoi*, of things. Pliny said, speaking of first causes: *Latent in majestate mundi*, — "They are hidden in the majesty of the world." The thought is as fine as its expression is eloquent. Let us leave, then, to Nature her secrets; and, if we are led to believe that the Sun pours on the Earth and the planets animated germs, refrain from trying to penetrate farther the essence of this mysterious phenomenon. We ask not of the stone why it falls, of the Earth why it turns, of the tree why it grows, of our heart why it beats, of the Sun's rays why they produce life on the Earth and immortality in the heavens.

CHAPTER XXIII.

Practical Rules derived from the Facts and Principles developed in this Work. To ennoble the Soul by the Practice of Virtue, by seeking Knowledge by Science of Nature and her Laws. To render Public Worship to Divinity. The Imperfections of Actual Religions founded Four Thousand Years ago in a Time of Ignorance and Barbarism. The Religion of the Future will be based on Science and Knowledge of the Universe. The Memory of the Dead should be preserved. We ought not to fear Death. It is but an Imperceptible Transition from one State to Another,—not an End, but a Metamorphosis. Impressions of the Dying. Whom the Gods love die young.

WE will conclude by setting forth the practical rules that result from the facts and principles expounded in this work.

Since man cannot rise to the superhuman in rank until his soul has been sufficiently purified, it evidently behooves him to care for the culture of his soul, to free it from every stain, and save it from every fall. Be good, generous, compassionate, grateful, approachable by the weak, friendly to the oppressed. Comfort those who suffer and those who weep. Practise charity in all its forms. Love to raise your thoughts above earthly things. Resist the material propensities which are the stamp, the stigma of terrestrial life. Aspire to the good and the beautiful. Live in the highest spheres, the farthest removed from earthly connections. Only at this price can you exalt and ennoble your soul, and fit it to enjoy the better life that awaits it in the ethereal domains, and to put on the new garment: this will be its passport to new horizons beneath the deepest heavens. For if your soul is vicious and corrupt; if during your entire earthly life you have remained sunk in material interests, wholly given up to purely physical occupations, and courses which confound you with animals; if your heart has been hard, your conscience mute, and your instincts low and wicked, — you will be condemned to begin a second life on Earth. You will once, or several times more, drag along the burden of life on this ill-fated globe, where physical suffering and moral evil have their chosen homes, where happiness is unknown, and misery the universal law.

There is another reason for the careful cultivation of the faculties of our souls, and for incessantly purifying ourselves by the culture of the good. The noble and high-minded, the choicest souls, are the only ones, we have said, who are fit to communicate with the dead, with the loved ones they have lost. If, therefore, we are stained with moral unworthiness, we shall receive no tidings, no succor from the beings whom we have loved, and who have left us. Here is a powerful motive for incessant self-improvement.

One of the best means of ennobling and improving the soul, of raising it above earthly conditions, of bringing it near to the sublime spheres, is science. Study, strive to know Nature, to understand the phenomena and the media which surround you, to explain to yourself the Universe of which you are a part, and your soul will expand. It is truly sad to see the shameful ignorance in which lives almost all mankind. The population of our globe numbers thirteen hundred millions: of this number, scarcely ten millions could be found who have studied the sciences, and whose minds are really cultivated. All the rest are given over to an intellectual passiveness, which almost identifies them with animals. The Earth is but a vast field of ignorance. With respect to knowledge, nearly all men die as they were born: they have added not a single idea or acquirement to those that their parents, themselves ignorant, inculcated in them in their tender years.

Yet, thanks to the studies and vigils of a few men, the knowledge that we now possess is immense: we have taken gigantic strides in the study of Nature and her laws.

We know the mechanism and the order of the Universe: we have learned to reject the untrustworthy evidence of our senses, and have discerned the real course of the different stars, so alike in appearance, which shine in the firmament at night. We know that the Sun is fixed in the centre of our world; and that a retinue of planets, of which the Earth is one, revolve around it, in an orbit whose mathematical curve has been exactly determined. We know the cause of days and nights, as well as of the seasons; and we can predict, almost to a second, the return of the stars to a certain point in their orbits, their conjunctions, eclipses, and occultations. The globe on which we live has been traversed and explored with so much care that there is not one of its corners unknown to us. We know the cause of the winds and the rains, we can indicate the exact passage of the feeblest current of the seas, and we can foretell a long time in advance the hour and the height of tides throughout the globe. We know why there are glaciers at the two poles of the earth, and why other glaciers crown lofty heights. The movements of the ground which in past times produced ranges of mountains, and even now occasion volcanic eruptions and earthquakes, we have succeeded in explaining. The composition of all bodies on our soil, or hidden in its depths, has been determined with certainty. We know what contains air, and what constitutes water. There is not a mineral, not a rock, not a particle of earth, of which we cannot designate the true components. More than that, we can state the composition of the soil of the planets and their satellites, the stars that move over our heads at incalculable distances, and which only our eyes can reach. Science has wrought the miracle of chemically analyzing a body that it cannot touch, and which it sees only millions of leagues away.

We have studied, classed, and named all the living beings, animals and plants, that people the Earth.

There is not an insect hidden in the grass of the prairies which has not been described, and set in its true place in creation: there is not a blade of grass that the pencil of the naturalist has not copied.

Still more, Science has penetrated beyond the reach of our eyes. She has invented a wonderful instrument which has unveiled to our ravished gaze a whole world, whose existence would never have been suspected without its aid. The world thus revealed to us is that of the "infinitely littles." We know that in a drop of water there are myriads of living beings, animals or plants; and that these creatures, so infinitely small, have life as complete and well organized as their full-grown analogues, and that the physiological functions of all these imperceptible creatures are discharged as well as ours.

Just as we have penetrated the life of the "infinitely little," we can pierce the depths of celestial space, and scan with our eyes the magnified image of stars that move at immense distances from us. The telescope displays to our view the surface of the moon, showing it seamed with deep ravines and rough unevennesses, bristling with enormous mountains, which are furrowed with yawning, circular crevices. We can traverse with our eyes the lunar disk, as if it were a far country of our own globe. Even of the planets which disappear in the infinity of the heavens, we can, thanks to the magnifying power of telescopes, judge of their aspect and their surface.

According to this statement, incomplete though it is, of what human science has been able to gather, one would suppose every inhabitant of the Earth to be not only eager to appropriate this knowledge, but also to be happy and proud in storing his mind with it. Alas! almost the whole human race knows not the first word of all this. Leaving out the ten millions of individuals of whom we spoke just now, — and who, numerically, are of little account, compared with the population of the globe, — all mankind imagine that the Earth is a plane surface extending to the limits of the horizon, and crowned with a blue cupola called the heavens. If you assure them that the Earth turns round, they begin to laugh; and they cite the motionless Earth, and the Sun that *rises* at the *right* and *sets* at the *left*, as proof positive that the Sun goes and comes. Poets say in real earnest that the Sun leaves his bed in the morning and retires to it at night! With the poets the people believe that the stars that shine at night in the celestial vault are ornaments, a pleasant spectacle intended to please our eyes, and that the moon is a cheap lantern. Nobody troubles himself about the cause of rain and fine weather, of cold and heat, of wind and tides. Every one shuts his eyes to these phenomena, for fear of having the trouble of explaining them. If a curious and ingenuous child asks his father to explain the simplest fact, — the cause of rain, of snow, of dew, — the father makes some senseless answer, or changes the subject, not knowing what to say. Nature is a sealed letter to most men, who live amid the most curious and varied phenomena, like a horse who has his eyes covered with blinders, so that he can see only straight ahead, or like a miner who works at the bottom of his shaft, seeing nothing but his tools and his task. Man's attention is occupied in eating and drinking, and devising ways of injuring his fellow-men.

Yes: it is a sad spectacle, — humanity thus preoccupied exclusively with material needs, and utterly indifferent as to all intellectual labor! And it is painful to think that such is the condition of nearly all the inhabitants of the globe. How superior to the mass of his fellow-men is he who has known how to cultivate his mind, to enrich it with sober and useful knowledge, and appropriate to himself one branch of the diversified tree of exact sciences. What power and grasp must his soul, thus fortified, have acquired! Strive, then, reader, to study and learn. Acquaint yourself with the secrets of Nature, account for every thing around you, comprehend the Universe and its innumerable productions, admire the power of God in thoroughly understanding His works. Then you will not go to the tomb with a soul as vacant as it was at your birth. In the last hour of death you will be *sage*, — a word that, according to Latin etymology, means wise (*sapiens*); and, feeling yourself nearer the sublime nature of superhumans, you will be fitted to soar by their side in the ethereal spheres.

To exalt and improve the soul, it is necessary not only to put virtues into practice, and to inform ourselves: we must strive to understand and love God, the Author of the Universe. Men, enter the temple, and bow down before God, according to the forms and rites of worship in which your youth was trained. All religions are good and worthy of respect, because they enable us to render to God the homage of grateful and submissive hearts. Christianity is good, because it is a religion. Mahometanism is good, because it is a religion. Buddhism is good, because it is a religion. Judaism is good, because it is a religion. The religion of the savage tribes of America, who worship the Sun in the depths of their forests, is good, because it is a religion.

In every religion there are doctrine and worship. The question now is of worship. In fact, in all modern religions, the form of worship is well conceived: it harmonizes with the habits and customs, with the dash of imagination and poetry, in every people; so that external manifestations correspond with the traditions and spirit of each country. As to doctrine, it is another thing. In the different religions to which, to-day, all the nations of the Earth adhere, doctrine is decaying and decrepid: it cannot endure the scrutiny of reason. The doctrine of Buddhism, which restricts human life to the earthly existence, which denies personal immortality to man, — absorbing the individual, after his death, in the bosom of the Great All, — is monstrous immorality, revolting pantheism. The doctrine of Mahometanism — which has no basis but the words of its founder, gathered under the title of the Koran, and regarded as divine revelation — is not taken in earnest by the Mussulmans themselves. The doctrine of

Judaism, which rests on the advent — always vainly expected — of a Saviour Messiah, the need of whom is in no wise apparent, is almost ridiculous. The doctrine of original sin, which lies at the foundation of Christianity, is illogical and unjust. To hold all human kind, present and future, responsible for an alleged infamy of our race; to make it pay the penalty of a wrong done, six thousand years ago, by one man and one woman, in an obscure corner of Asia; to say that God had a Son, and that he sent this Son to ransom all men condemned and lost in consequence of Adam's sin, — is contrary to reason.

A deplorable fatality would have religion — that supreme want of souls, that powerful element in the moral elevation of the masses, that precious means held out to enlightened and reasoning man to draw near by thought to the Divine Author of the Universe — rest to-day, among all the inhabitants of the two hemispheres, on very inaccurate bases.

The trouble is, that all religions were formulated, in their essential doctrines, four thousand years ago, when the thickest darkness surrounded the human mind. In this infancy of civilization, men could have only conceptions proportioned to their feeble knowledge. They made a God in their own image: they gave him their mean passions, — jealousy, hate, revenge, dissimulation, anger. More than this, they gave him their own form: they made God a handsome old man with a white beard! Fontenelle wittily said that if God made man in His own image, man had paid Him well for it.

When religions were founded, nothing was known of all the Universe but the Earth, and only a small part of that. The moon and stars were believed to be little lamps, hung in the heavenly vault to light up our firmament and mitigate the darkness of the nights. The Sun was a torch for illuminating the Earth, and which lit nothing but the Earth. The other stars that make up the Universe were nothing to the ancients, who did not even suspect the existence of other worlds. The Earth was all: it was, in itself, the world.

Conceived in such absolute ignorance of the Universe, religions must necessarily go to the wall whenever the true order of the world was made known, when the boundless immensity of the Universe was appreciated, and when it was understood that the Earth is but a dot in space, and plays among the heavenly bodies but a very subordinate part. This happened in 1610, when, for the first time, the telescope (then just invented) was directed upon the moon by Galileo. That did the business.

Eminent prelates, the enlightened men who then controlled the destinies of the Roman Church, were not deceived. The cardinals and all the Holy College saw instantly the dangers that threatened them; and their course proves that they clearly understood that the discoveries of astronomy were going to shake and throw down the edifice of existing religions. Scarcely had the gleam of scientific light become visible, when the hands of the Church set about quenching it. Rome declared war to the death against the new astronomy. There was Pierre D'Albano, author of a treatise on astronomy, burned in effigy at Bologna, in 1327; and Cecco d'Astoli, given to the flames, at Florence, in the same year, for having proclaimed that the Earth moved. There was Jordano Brieno, who mounted the blazing pile, at Rome, Feb. 17, 1600, his crime being a profession of the same belief. There was the naturalist Antonio de Dominis, whose decayed remains were disinterred in 1625, at the Château Saint Ange, where he had died a prisoner, in order to be cast into the avenging flames. There was Campanella, seven times tortured and twenty-seven years a prisoner, because he yielded assent to Galileo's philosophy.

Copernicus—a canon, alas!—could not be persecuted by the Romish Inquisition, for the sufficient reason that he died before the publication of his book, the "New Astronomy,"—privileged only to touch with feeble hands, on his dying-bed, the very first copy that fell from the press. His enemies revenged themselves for not having burned him dead, by burning his book, the first cause of the rebellion. Kepler, the immortal continuer of Copernicus's work, a Protestant, and never quitting Protestant England, was nevertheless followed throughout his life by the hate of the minions of the Church. He was accused of heresy. His aunt was burned for sorcery at Weil. His mother, also accused of sorcery, was imprisoned at Stuttgardt, in 1615. She remained in confinement five years, and was saved only by the wonderful devotedness and unceasing labors of her tender and unhappy son. Kepler himself lived, moreover, the most unquiet and troubled life that a man of genius could endure.

There was Roger Bacon, the learned friar of Oxford, who, leading his age by his scientific discoveries and by the persecutions which they drew upon their author, passed the greater part of his life in prison, — now in a cell of his convent, now in a dungeon. His crime was the study of physics and astronomy. Two centuries later, his namesake Francis Bacon was enrolled on the "Index Expurgatorius," by the English ecclesiastical authorities, for the same reason.

In France, our illustrious Descartes was a wanderer and exile throughout his life. He was pursued everywhere by the hate of bigots, — Descartes, the religious man, the spiritual philosopher *par excellence*, whose orthodoxy was profound and sincere, who never pronounced and never heard the name of God without removing his hat in token of respect. But he was an astronomer, and for that reason deemed an enemy of the Church.

There was the learned Jesuit, Fabri, imprisoned at Rome, for saying in a sermon that, "the motion of the Earth once demonstrated, the Church must interpret in a figurative sense those passages of Scripture that are opposed to that principle."

At the same period — that is, about 1630 — Galileo was bitterly persecuted by Romish prelates. The publication of his immortal Dialogues, in which the doctrine of the Earth's motion was proved, together with other new truths in astronomy and physics, excited a storm of rage, which ceased only at the death of the learned but ill-starred Florentine. And what proves that Pope Urban VIII. did not, as has been alleged, persecute Galileo in mere petty and personal spite, but because he wished to protect or avenge the Christian religion, is the correspondence between those around the pope, and Galileo, during his trial.

"Put a close guard on your words," Ciampoli wrote to him, in February, 1615; "for where you merely establish a certain resemblance between the Earth and the moon, another goes farther, and says that you believe that there are men living in the moon. And this other begins to argue that these men cannot be descendants of Adam, or have come out of Noah's ark."

His solemn abjuration and his humble attitude before the judges of the Holy College could scarcely save Galileo from the flames. The pile which had consumed the body of Antonio de Dominis, in 1625, still smoked on the field of Flora; and from the church of the Convent of Saint Minerva, where he pronounced his abjuration, the unhappy old man could see the theatre of his posthumous punishment.* The rest of his life Galileo passed in a partial captivity and perpetual exile, at his country house of Arcetri, where all freedom of action was denied him. The Church could

* See the history of the trial of Galileo, in the author's "Vie des Savants Illustres," Book IV.

not forgive this great man the deadly and irreparable blow that he had inflicted upon her by popularizing the new astronomy.

In France, also, the Catholics clearly saw that astronomical discoveries must overthrow the old theology. Gassendi, professor in the College of France, and the scientific oracle of his time, believed in the motion of the Earth; and Le Cazre, rector of the College of Dijon, sought to divert him from this opinion, by showing him its theological consequences. He wrote to Gassendi: —

"Think less of what you yourself may believe than of what most other men will think, who, drawn by your authority or arguments, will persuade themselves that the Earth moves among the planets. They will conclude at first that, if the Earth is beyond question one of the planets, since it has inhabitants, it is easy to believe that the others have them; and that they are even in the fixed stars; that they are of a superior nature, in proportion as the other stars surpass the Earth in size and in degrees of perfection. Hence will arise doubts about Genesis, which declares that the Earth was made before the stars; which were not created until the fourth day, to light the Earth and measure its seasons and years. In consequence, the whole economy of the incarnate Word and evangelical truth will be looked upon with suspicion.

"What do I say? So it would be with the whole Christian faith; which believes and teaches that all the stars were made by the Creator not for the habitation of other men or other creatures, but only to illuminate and fecundate the Earth with their light. You see, then, how dangerous may be the spread of these things among the people; especially among those who, by their authority, seem to make a declaration of faith of them. Not without reason did the Church, in the time of Copernicus, always set its face against this error; and, still later, not certain cardinals, as you say, but *the supreme head of the Church by a pontifical decree condemned it in Galileo,* and in the most solemn manner forbade its teaching in the future, by word of mouth or by writing."

Vain efforts! Ideas born in a period of darkness and ignorance could not live in a time of light, and contemporary with a knowledge of the actual world. They had to vanish under the new illuminations of science. To-day mankind reasons. Religious doctrines, in order to be accepted, must be based on the true order of this boundless Universe that the ancient religions ignored. The human race must be regarded no longer as the centre or objective of all Nature, nor as dominating all that is visible under the skies, but rather as a mere particle of creation, an obscure member of the general family of worlds. Far from affirming that every thing was made for man, it should be declared that the Universe is a continual whole, — an unbroken chain, of which mankind is but a link. It must be understood that the Earth is only a grain of sand, lost in the immeasurable extent of infinite space.

See on what positive foundations the religion of Science and Nature will be built. This new religion will be the work of the twentieth century. Then, the minds of men being better prepared than they are to-day for this moral revolution, the new doctrines will be easily accepted. They will entail no struggles or combats. While the old religions have arisen and grown great in blood and tears, by persecutions and torments, amid the sufferings of martyrs and cruel repressions of the adherents to old doctrines, the religion of the future, prepared by unanimous consent, by universal conversion, will rise at the cost of no tear, no drop of blood. It will spread rapidly over the whole Earth. Then, the extreme facility of communication having distributed these simple and true ideas in the various parts of the world, gradually all the peoples will adopt the new religion. Its advantages and its conformity with the order of Nature will be so probable, that every nation of the two worlds will embrace it, as each will have adopted, after having recognized its advantages, uniformity of weights, measures, and money, founded on the metric system.

But we are not yet in the twentieth century. We are still in the nineteenth. We are not in the presence of the religion of Science and Nature, but in the presence of many and diverse religions, all imperfect in doctrine, but admirable in point of public worship. Let us attach ourselves, therefore, to this worship, which is the only means of establishing our relations with Divinity, and of keeping in our hearts the idea of the Supreme Being. Catholics, go into your churches, and amid the superb pomp of your holy ceremonies lift your grateful souls to God, humble yourselves before the Sovereign Ruler of the Universe. Protestants, in your temples, sing your psalms and canticles. Russians and Greeks, kneel in meditation before your dazzling and mysterious tabernacles. Jews, frequent your majestic synagogues, cherish the perfumes which, addressing the subtlest senses, speak of God to softened souls. Mussulmans, repair to your tranquil mosques.

Buddhists, learn the way to your pagodas. Savages of two worlds, who adore the Sun in woody solitudes, lift your comforted hearts to the radiant star. Let all men, in every land, under all skies, practise the religion in which it was their lot to be born, until the growth of reason in the popular mind has helped to create the religion of Science and Nature. All is good, all is beautiful, that helps us to render homage to Divinity. Religious worship is the prime need of our souls, as it is the guaranty of peace and happiness in society.

The fourth practical rule, that we lay down as a corollary of the principles and theories that we have set forth, is to remember and tenderly regard the dead. Let us not efface from our hearts the memory of those whom death has taken from us. To forget them is to subject them to the most cruel anguish, and to deprive ourselves of the succor and support that they can give to guide us here below.

The ancients carefully cherished the memory of the dead. They did not shrink in terror, as moderns do, from the idea of death. On the contrary, they loved to invoke it. Among the Romans and the Greeks, cemeteries were places of reunion, used for promenades and festival enjoyments. The Orientals of our day have preserved this tradition of antiquity. Their cemeteries are gardens assiduously preserved, in which the cheerful crowd saunters every festal day. They pay visits to their relatives and friends buried beneath shrubbery and clumps of flowers. They give themselves up to the pleasures of life in these joyous asylums of death.

In our Europe we are wide apart from this custom, which was inspired by a healthful philosophy. Only, it should be remarked, peasants, nearer to Nature than are dwellers in cities, are far from shrinking from the idea of death and shunning the places where sleep their friends and relatives. Rustics love to evoke the memory of the dead. They speak of them, talk to them, consult them, just as if they were sitting around the family fireside.

The custom of funeral feasts, which dates back to primitive man, is preserved in many countries. Returning from the cemetery, the mourners sit down to a well-laden table in the house of the deceased, and wish him a happy journey to the land of shades. In our cities, it is mostly the masses who deem it a duty to bring flowers to the graves of their relatives. Members of the higher classes are generally exempted from this pious care; and this is wrong. Piety toward the dead, and the preservation of their memory, are prescribed by the laws of Nature.

We will say to the reader, as the consequence and final practical rule resulting from what goes before, that he ought not to fear death. Let him face that moment, so dreaded by all men, with a steady and quiet eye. Death, we have said, is not the end, but a change: we do not perish, we are transformed. The caterpillar, that seems to die in shutting itself within a cold grave, does not die. It is soon resuscitated in the guise of a brilliant butterfly which flits through the air. The ternal, motionless, and cold body of the chrysalis gives place to a glittering shape, variegated in a thousand hues, and cutting space with its azure wings. It will be thus with us. If our wretched tenement remains on the Earth, and restores its elements to the common reservoir of universal matter, yet our soul will not perish. It will be born again, an invisible butterfly, which will traverse the air and soar in the ethereal regions. It will quit our Earth, where pain and evil are the constant law, for a blest home where all the conditions of happiness are united. Why, then, should you fear death? If you cannot long for it, you must at least await its coming hopefully and calmly. Death will reunite us with those we have loved, those whom we love and ever shall love. What a deep spring of consolation is this for the remainder of our life! What a store of courage for the dreaded minute of our own death! Dear and gentle departed, you who have never ceased to be held in memory, your loss has rendered us — at the price, it is true, of the cruelest anguish — the sorrowful and supreme kindness of soothing the severities of our future agony. The sadness of our last moments will be calmed by the thought that you await our coming; that you are ready to receive us at the threshold of the other life; that you are going to be our guides in the new domain which opens to us beyond the grave!

The fear of death, which chills the hearts of most men, seems to lose much of its weight at the last moment. Those who professionally attend the dead, like priests of different religions, physicians, attendants in infirmaries, and Sisters of Charity, know that most death-struggles are easy. Whoever dies at the end of a noble and honorable life feels at this solemn moment that he is going to a new and better world. He is happy, and his happiness betrays itself in his words and the expression of his eyes. The only thought that saddens him is of the pain that his death is going to inflict upon those whom he loves and is about to leave. Says Montaigne: —

> "I believe, indeed, that the ruins and dreadful apparatus with which we surround death, cause us more fear than death itself. A form wholly strange, the cries of mothers, women, and children, the visitation of persons astonished and overcome, the presence of pale and weeping servants, the darkened chamber, burning tapers, the bed besieged by physicians and preachers, — sum total, all horror and alarm around us. We seem to be already buried and under ground. Children fear their own friends in masks. So do we. This mask should be taken off things as well as off persons."

Those who have watched the dying have made observations which we will state summarily.

First, we must leave out of such observations deaths occasioned by maladies that destroy the consciousness of the dying. Such cases are very many. Think, for instance, of deaths caused by cerebral or pulmonary apoplexy, by rupture of aneurism, or affections of the heart, which entail speedily fatal symptoms. In all these cases, the organs of speech being paralyzed, the dying can express nothing. To learn the thoughts of the dying, we must observe those who, up to their latest breath, preserve their intellectual powers unabated, — who "have their head," as the saying is. It is certain that their dying struggles are very tranquil. Consumptives, wounded persons, those dying from disease of the stomach or the intestinal canal, or of those fevers that sap the strength without affecting the mental faculties, the dysenteric and the dropsical, who retain to the last minute full possession of their intelligence, die calmly and almost with delight.* M. de ——, Captain of Franc-tireurs, in the Vosges, who, in a fight with the Prussians, was struck by a bursting shell in the abdomen, and died a few hours later, said, as he expired, "What happiness! I am going to see my dear wife again."

It should be added that in most cases death has been preceded by a gradual prostration of strength or sensibility, so that the dying person is hardly conscious of the change that is about to take place in himself, and he faces the moment of death with absolute indifference. Buffon has well stated this fact in his chapter on Man: —

"Death, that change of condition so marked and so dreaded, is, in Nature, only the last shade of a preceding condition. The necessary succession of decay in our bodies brings on this stage, as all others that preceded it. Life begins to wane long before it is utterly extinguished, and in reality it is farther from decay to youth than from decrepitude to death; for we must regard life here not as something absolute, but as a quantity susceptible of augmentation or diminution.

"Why, then, should we fear death if we have lived well enough to feel no apprehensions as to what follows it? Why dread that

---

* The reader may consult on this subject a book called "De l'Agonie et de la Mort," by H. Lauvergne. 2 vols. Paris, 1842. It is the work of a cool and methodical observer, — a physician, — devoid of enthusiasm; but its facts are numerous and well arranged.

moment, since it is prepared by an infinity of other moments of the same kind, since death is as natural as life, and both happen to us in the same way, ourselves not being sensible of them, or being able to perceive them. Ask physicians and the ministers of the Church, accustomed to observe the actions of the dying and to receive their last sentiments They will agree that, except in a very small number of acute diseases, where the agitation caused by convulsive movements seems to indicate pain, the dying pass away quietly, gently, and painlessly. Most men die without knowing it; and of the few who retain consciousness up to the last sigh, there is not one who does not retain hope as long. . . .

"Death, then, is not so terrible a thing as we imagine it to be. It is a spectre that frightens us at a distance, and disappears as we approach it." *

There is surely a time that often lasts several hours, and in which, life having wholly withdrawn from the body, it is already a corpse under the eyes of those present; and this corpse still moves and speaks. But the soul that survives in this body already cold and actually dead is not that of a terrestrial man: it is already a superhuman's. The dying man has consciousness, and even perhaps an anticipative sight, of the ineffable bliss that awaits him in the new world whose threshold he is touching; and he manifests his joy in speech, and in the expression of his eyes. His last sigh passes in a flight of supreme joy.

This extraordinary state in which the dying are half on earth and half in the new realm to which they are destined, — having, so to speak, one foot on earth and the other in heaven, — accounts for the touching eloquence, the often sublime words, that flow from their failing

---

* Natural History of Old Age and Death. Book. II. p. 579.

lips. An ignorant and uncultivated man expresses himself on his death-bed with an eloquence unaccountable to those who hear it. In this way are explained the prophesies of the dying that subsequent events have verified. The dying have an insight into facts of which they would not have had the least notion, if they shared the common conditions of human kind. For this reason we should treasure their last words with religious care, and scrupulously regard the wishes they express.

In Moldavia, when a peasant has escaped from a severe illness, in which he has seemed to touch the very portal of the tomb, his friends press around his bed to ask what he saw in the other world, and to get news of their relatives gone before; and the poor sick man tells them his visions as well as he can.

Constant Savy, a modern writer who left some tracts on spiritual philosophy, in his "Thoughts and Meditations," related an extraordinary dream that he had at a time when death seemed imminent. We will quote this strange and interesting narrative:—

"I was very ill. My strength was gone: it seemed to me that life attempted to resist death, but in vain, and was about to vanish. My soul gradually detached itself from the matter diffused throughout my body: I felt it withdrawing from all those parts with which it is so intimately connected, and gathering, as it were, at a single point, the heart; and a thousand thoughts, obscure and cloudy, about my future life, took possession of me. Little by little Nature was fading out before me, and taking under my eyes irregular and strange forms. I nearly lost the power of thinking. I retained only that of feeling. This feeling was all love, — the love of God, and of those whom I had held most dear in it, but powerless to manifest that love: my soul, concentrated on a single point of my body, had ceased nearly all connection with, and could no longer rule it. It still felt some distractions, however, caused by the sufferings of this body, and by those which surrounded me. My life was bound to matter by only one of the thousand threads which once united it thereto: I was about to expire.

"Soon, to mark, no doubt, the passage from this life to the other, there came something like thick darkness, followed by a brilliant light. Now, O my God! I see Thy day,—that day so longed for! I see together, full of joy, those whom I had so loved, who had inspired me in my bereft life on Earth, and seemed to dwell in my soul or to hover over me. They awaited me; they welcomed me with delight. It seemed as if I filled out their lives, and they filled out mine. But how different were my sensations of happiness from those in the life I had left! I cannot describe them. They were penetrating, yet not vehement; they were gentle, tranquil, full, unalloyed, without a void, without unrest, ravishing, ineffable; and yet they were joined to the hope of the greatest happiness. . . .

"I see Thee not, O my God! Who can see Thee? But I loved Thee more than I had loved Thee in this world. I understood Thee better, I felt Thee more powerfully. Thy foot-prints, which are visible everywhere and in all things, seemed more palpable and splendid. I felt admiration and astonishment such as my soul had never known; I saw more clearly a part of the wonders of Thy creation. The bowels of the Earth had no more secrets from me: I saw them in every particle,—I saw insects and other creatures that live therein, the quarries that form the frame-work of the globe, the mines known to man, and those unknown. I read its age in its bosom as one reads the age of a tree in its heart; I saw all the conduits that bear to the sea the waters that sustain it; I saw the reflux of these waters like the course of blood in the human body, from the heart to the extremities, from the extremities to the heart. I saw the depths of volcanoes; I understood the trembling of the globe and its connection with the stars; and, as if this globe were turned in every direction to exhibit itself to me and make me admire Thy grandeur, O my God! I saw all lands, with their diverse populations and their different customs; I saw all the varieties of my kind, and a voice said to me, 'Like thee, all these men are in the image of the Creator; like thee, all will march steadily toward God, conscious of their progress.' The density of the forests, the depths of the seas, could hide nothing from my eyes: I could see all, admire all, and I was happy in my own happiness and in that of the dear objects of my fond love. Our joys were mutual. We felt bound together both by our old affections grown far deeper, and by the love of God. We drew the same bliss from the same spring; we were but one; we enjoyed together and separately this inexpressible felicity. I am mute that I may feel it more deeply."

Without going to the farthest limit of the death pang, it is easy to convince ourselves that those who are doomed by Nature to an early death, those who must die young, possess a deep serenity of spirit. This moral appanage is, in our opinion, one proof that they have already a presentiment, or even the anticipative enjoyment, of the new life that awaits them after death. Why have consumptives such sweetness of temper, such quick sensibility, hearts so expansive and susceptible that everybody notices these peculiarities, characters so marked as to aid the physician in making a diagnosis of their disease? It is, we think, because these sick persons, already half-gone from the Earth, have already partially taken on the moral attributes of superhumans. Consumptives, it is well known, are always confident of recovery; they lay plans for enjoyment and the future, when their last hour is about to strike; they feel hope and joy, while friends are thinking of their funerals. It is commonly said, in explanation of this anomaly, that consumptives do not appreciate the gravity of their disease: for our part, we think that they have, on the contrary, some confused and dim idea of their condition; we believe that Nature reveals to them the approach of a life of unclouded happiness, and that it is this secret conviction that gives them hope and confidence for the future. The future that they catch a glimpse of is not that of earth, but that of heaven.

Alexandre Dumas, the younger, has aptly expressed this truth in a beautiful page of his romance "Antonine," which we may be permitted to quote: —

"Did you ever know consumptives to be aware that they were such? Have you noticed that for them life has aspects unknown to those who have much longer to live? Their eyes, to which, by the presentiment of death, God partly unveils His eternity, see beings and objects in a peculiar and poetic light. They see with their spiritual rather than with their physical vision. In them sensations are electrically instantaneous, — what moves others only through deduction moves them at first sight. One would say that their souls, too closely cramped in their breasts, strive constantly to rise; and that, from the heights which they reach, they discern what escapes the common eye. Their souls live higher than their bodies; and this accounts for their easy death; for, when the last hour comes, their immaterial part has been so long separated from its corporeal envelope, that it easily and painlessly detaches itself from and abandons it, as we cast off a garment that is too heavy . . .

"Those who are attacked with this disease have, like the sick man of Milleroye, who was no other than Milleroye himself, an incessant longing to draw near to Nature, the first source of life. For them the trees have a peculiar shade, the birds sing songs that only they can understand, the sun dispenses a heat that others feel not. Where others see nothing but a natural fact, they see a blessing from God. Their faces at last take on the sad poetry of their spirits. For suffering they feel the very pity that they inspire. They are charitable, and forgiveness is habitual in them because they are near the Lord. If Nature has granted them the power of reproducing in bodily expression the sensations that life awakens in them, their talent suddenly becomes genius, it wears a pale and transparent hue like a star-ray, and exhales a perfume like the fragrance of a hidden flower. Hear Bellini, read Milleroye; and you will find, in the music of the one and the verses of the other, that indefinable sentiment, plaintive and melodious, which has been their very life."

Not among consumptives alone may these observations be made. Every man predestined to die young seems marked with that secret sign of the soul which produces sometimes a sweet and charming melancholy, and again a vivacity or sensibility that relatives admire, and that is, alas! too often the signal of approaching death. The beautiful qualities that shine in these young people are but the forerunning indices of their dissolution.

"Short lived are children born with such great minds,"

says Casimir Delavigne in "The Children of Edward." *

The Greeks said, "Those who die young are loved by the gods."

Therefore let us not fear death: let us await it not as the end of life, but as its transformation. Let us learn, by the purity of our lives, by our virtues, by the cultivation of our faculties, by knowledge, by practising the worship of our ancestors, to prepare ourselves for the critical moment of that natural change which will bring us into the blessed mansions of the ethereal spheres on the To-morrow of our Death!

---

* Shakespeare has it, —

"So wise, so young, they say, do not live long." — Tr.

EPILOGUE.

IN WHICH GOD IS LOOKED FOR, AND IN WHICH, THE WAY HAVING BEEN MADE, THE UNIVERSE IS DESCRIBED.

THE author begs permission to record here a grave conversation that he held with his friend Theophilus to whom he had shown, in order to obtain his advice and opinion about it, the manuscript of "The To-morrow of Death." The words of this conversation fill a hiatus which existed in this work, as will be seen. Let us allow the two interlocutors to express themselves in the ordinary fashion of dialogue.

THEOPHILUS (*entering the author's study, and throwing the manuscript on the table*).

I have read your manuscript from beginning to end, and I will tell you soon my impressions as to matters of detail; but, first, let me call your attention to a great gap in this work.

THE AUTHOR.

What is lacking in it?

THEOPHILUS.

God.

THE AUTHOR.

But —

THEOPHILUS (*interrupting*).

You are about to say that you frequently mention that name; that the words "Providence," "the Author of Nature," "the Creator of the Universe," &c., have fallen many times from your pen. This is true; but it is quite as true that you do not go beyond this vague denomination, that you say nothing of the person of God, and that you do not seek out His place in the Universe that you run over with such great strides, in company with souls more or less spiritualized. Why this reserve? And since you tell us that souls wholly spiritualized inhabit the Sun, why do you not tell us where, according to your system, God, the Sovereign Master of these souls, sits enthroned? What reasons had you for ignoring a question of such great moment?

THE AUTHOR.

Reason enough. First, I have everybody's reasons. The idea that must be formed of God, to put Him in harmony with the boundless immensity of this Universe which came from His hands, so transcends the grasp of the human intelligence, so overwhelms our mind, that we stop, powerless, discouraged, and almost terrified by the audacity of daring to ask what God is.

THEOPHILUS.

Yet I should be surprised at your shrinking from such an enterprise. When a man has constructed a complete system of the Universe, he does not stop on the road; and I can hardly believe that, having dared to establish on each step of the ladder of your theory all the elements of the solar world, planets and their satellites, stars and asteroids, plants, man, and animals, visible and invisible beings, souls and bodies, matter and spirit, you have not indicated the Creator's place; that in this mighty edifice of the world you have classed every thing except the Sovereign Architect.

THE AUTHOR.

You are right, my friend. God has His place in my system.

THEOPHILUS.

Then why don't you say so? Why are you silent on this point?

THE AUTHOR.

There are bold statements enough in my book already: I expose myself so amiably to the rage of materialists and the hate of bigots, to the joint animosity of the wise and the ignorant, that I fear to give them another pretext, another weapon for their castigations.

THEOPHILUS.

That is no reason. Why do you take up your pen if you dread discussion and fear detraction? You could have abstained from it, and kept to yourself your ideas about the origin and destiny of man; but as soon as you submit them to the public, you must utter your entire thought. If you think your system is right, you must expound it without reserve.

THE AUTHOR.

My friend, Wisdom speaks through your lips. I should therefore simply bow, and follow your advice. . . . Yet I dare not do so at once. I will therefore propose to you a middle course, a *mezzo termine.* In confidence and in the freedom of this *tête-à-tête*, I am going to state to you my thoughts about God, to tell you in what part of the great Universe I place His dazzling personality. If the idea seems to you absurd and untenable, or merely rash, your frankness will not conceal it from me; and, duly advised, I will keep my theory to myself: otherwise —

THEOPHILUS (*interrupting*).

Admirable. Nothing can be said against that. I have only to listen to you. I do listen.

(Here friend Theophilus seats himself in a comfortable chair, places an ottoman under his feet, a book under his elbow to support it, and a cigarette of Turkish tobacco between his lips, and sets himself to the task of listening with a grave air of collectedness, relieved by a certain touch of suspicious severity, as becomes the arbiter in a literary and philosophic matter.)

THE AUTHOR.

You wish to know, my dear Theophilus, where I locate * God? I locate Him in the centre of the Uni-

---

* This word, however objectionable on general principles, seems to express the author's meaning better than any other English verb. — TR.

verse, or, in better phrase, at the central focus, which must exist somewhere, of all the stars that make the Universe, and which, borne onward in a common movement, gravitate together around this focus.

THEOPHILUS.

Pardon me. I do not exactly catch —

THE AUTHOR.

You will comprehend my thought better by and by. Remember only, to begin with, that I locate God at the common focus of the stars of the whole Universe. But where is the common focus? To know that, we must first understand the Universe, and the whole order of its movements.

THEOPHILUS.

That is explained in the course of your work.

THE AUTHOR.

You are mistaken, friend. I have described only the solar system in my work, and he would have a very incomplete and inadequate idea of the Universe who stopped at that. The *world* and the *Universe* must not be confounded, as is too often done. The *world* is our world, — that is, the solar system of which we are a part: the *Universe* is the union of all the worlds, or systems like our solar system. In the manuscript that you have just read, I was able to make known only a little corner, an insignificant fraction, of the Universe.

THEOPHILUS.

You call the solar world a little corner! That world, with Neptune which circulates eleven hundred and fifty millions of leagues from the Sun; Uranus, which rolls at a distance of seven hundred and thirty-two millions; Saturn, at three hundred and sixty-four millions; and in which there are comets that traverse thousands of millions of leagues, like that of 1811 which occupied three hundred years in passing through its orbit, or that of 1680 which requires eighty-eight centuries to complete its prodigious revolution, — the first of these comets wandering more than thirteen thousand million leagues from the Sun, the second thirty-two hundred thousand million!

THE AUTHOR.

Yes, my friend, our entire solar system, — the Sun, with its great retinue of planets and asteroids, with their satellites, with the comets that now and then drop into the glowing furnace of the mighty Sun, — all this, compared to the Universe, is but a grain of wheat in a full granary, but a particle of sand on the seashore, but a drop of water in the ocean, but an atom of floating dust in the whole mass of the air. So tremendous are the dimensions of the Universe, that it is absolutely unapproachable by our computations, and is to us the image of infinity, or infinity itself. Now, friend, mark this. Surely God is, in His nature, utterly inconceivable by our minds. His essence eludes and will always elude us. We can only affirm that He is infinite in His moral perfections and intellectual power. But if, on the one hand, God is infinity in the moral order, and, on the other, the Universe is infinity in the physical order; if the one is infinity in spirit, and the other infinity in extent, — these two ideas, however unapproachable in themselves by human intelligence, are yet of the same order, and may be brought together. We may, then, without incurring the charge of presumptuousness, locate the infinity which is called God in the infinity which is called the Universe; that is, establish the person of God in the common focus of the worlds that compose the Universe.

THEOPHILUS.

The reasoning is correct. It only remains to prove to — or, if you prefer it, to teach — me that the Universe is indeed infinity, by reason of its extent. I shall not admit this assumption except on very strong proofs.

THE AUTHOR.

Very well. Give me your attention, and excuse me if the demonstration that you demand resembles a lesson or a lecture on astronomy.

(*Here friend Theophilus sinks a little closer into his chair; he restores to its place the ottoman that had slipped from his feet, and, lighting a fresh cigarette, he leans to listen to his interlocutor, regarding him intently with wide eyes and half opening his mouth, which is the sign of concentrated, but at the same time friendly, attention.*)

THE AUTHOR (*resuming*).

I told you that our Sun, with all his retinue of planets, satellites, asteroids, and comets, that he draws and carries with himself through space, covering them with his vivifying rays, as a father carries his family, throwing over it his kindly protection, is only a corner of the Universe. You will understand this presently. When on a clear and calm night you contemplate the heavenly vault, you see it spangled with stars, whose number it would be absolutely impossible for you to count. But what you see with the naked eye is nothing. Take a good telescope, and look at a certain part of the heavens. Where you just now saw none, you will now see legions. You will see luminous points detached on the dark ground of space, like diamonds on the black velvet of a jewel-case. These brilliant points that the telescope reveals to you are stars just like those that we see at night in the celestial vault. I will now ask you if you really know what a star is?

THEOPHILUS.

Yes. I read in your manuscript, and I knew before, that the stars that we see twinkling in the heavens at night, and that we should see just as plainly in the day, were it not for the illumination of the heavens by the Sun, which pales their light to our eyes, — I know that these are stars that shine by their own light. Like our Sun, the stars are at once the centre of attraction and the torch of the particular world that they light up and attract toward themselves. Just as a whole battalion of planets, with their satellites, asteroids, and comets, move around our Sun, receiving from it heat, motion, and life, so the stars that are scattered through space impart motion and activity to a family of planets and satellites. These planets that move around stars constitute stellar worlds, analogous to our solar world. We cannot see the planets that attend these stars, on account of their diminutive size, and the immense distances that separate us from them, and which prevent us from discerning them, even with the most powerful telescopes: we see only the governing suns; that is, the stars. But the existence of these stars, fixed, like our Sun, implies the existence of planets moving around them.

THE AUTHOR.

Very well. So our solar world is not *sui generis:* it is only an individual member of a great family, the family of stellar worlds, which resemble our solar world in the arrangement and movements of the stars that they comprise. The Universe is composed of all these worlds united. You know all this, I see. But what, perhaps, you do not know — for we owe our knowledge of these facts to quite recent discoveries — is the great variety in arrangement and physical aspects that certain stars exhibit. In this respect, there is a sort of reversal of what constitutes Nature on our globe. Here we are in strange, improbable spheres. While maintaining their likeness to our solar world in the order of their movements, certain stars materially differ from it in the forces which direct Nature in them.

THEOPHILUS.

Explain yourself, I beg.

THE AUTHOR.

While a single central star governs our solar system, there are stellar systems governed by two suns; others by three, and even by four. In other words, there are stars coupled two together, or connected in threes or fours, and which act on the same family of planets, whose movements intermingle without damage. It is plain that the worlds that possess two or three calorific and illuminating centres must present, in their physical and mechanical conditions, peculiarities of which we have no idea.

There are other peculiarities of many stellar worlds. The light of our Sun, and that of most stars, is steady in its brilliancy: it neither waxes nor wanes. It is otherwise with many of the distant suns that we call stars. In some of them the light is seen alternately to grow feebler and stronger. Brilliant now, a little later they will be hardly perceptible: then we shall see them relume their fires, and shine as before. Some of them go out for ever. The philosopher Eratosthenes, 276 B.C., wrote, in speaking of the stars of the constellation Scorpio: "They are preceded by the most beautiful of all, the brilliant star of the *Serre-Borealis*." At this day, the "*Serre-Borealis*" is not the most brilliant in this constellation. Hipparchus, the Greek astronomer, 120 B.C., wrote: "The star in the fore-foot of Aries is exceedingly beautiful." This star is now not above the fourth magnitude. The first two stars in Hydra, which in the time of the astronomer Flamsteed, in the sixteenth century, were of the fourth magnitude, in the next century were of only the eighth. Like abatements of brilliancy in several stars have been observed by astronomers.

Stars that were formerly observed have disappeared in our day. One instance of many was a star which the astronomer Bayer had marked in his catalogue beneath the ε of the Great Bear, and which vanished from the heavens in the eighteenth century, according to Jean Cassini. In the constellation of Taurus, the eighth and ninth stars have also disappeared. The fifty-fifth star of Hercules began to pale in 1781, and in 1782 wholly disappeared. In 1437, the astronomer Ulregh-Beigh said that a star in Auriga, the eleventh in Lupus, and six stars marked in the catalogue of Ptolemy, the Egyptian astronomer, were not to be seen in his time. Stars have appeared suddenly, shed a very brilliant light, and soon vanished. Similar celestial phenomena are authenticated in the second century under the Emperor Adrian, and in the fourth under Honorius. In the second century a very bright star appeared in Cassiopeia. It surpassed in brilliancy Vega and Sirius, the most brilliant of all stars. On its first apparition it was visible in the day-time, as Sirius sometimes is, and as other stars are seen at the instant of a total eclipse of the Sun. But the new star soon began to wane; its light diminished more and more, and at the end of seventeen months not a trace of it was left. As the year 1572 had witnessed the terrible Massacre of Saint Bartholomew in France, the apparition in the same year of a new visitor in the heavens greatly exercised brains and tongues. In 1604 a new star appeared near the planet Saturn. In 1605 its brilliancy had materially diminished: in 1606 it was not to be seen. In the latter year another star, new risen, and which was called the Fox, presented a singular phenomenon. Before its final disappearance, its light was seen several times to fade and then brighten.

These augmentations and abatements of luminous brilliancy are not very unfrequent, moreover, in the stars that we know. The star *o* in the Whale varies greatly in its luminous intensity, and often disappears; the star χ in the Swan changes, under our gaze, from the fifth to the tenth magnitude; the thirtieth star of Hydra, which is of the fourth magnitude, disappears for nearly five hundred days. Other stars vary periodically in brilliancy, and these periods are very short: δ in Cepheus changes, in five days, from the third to the fifth magnitude; β in Lyra, in six days, from the third to the fifth; η in Antinoüs, in seven days, from the fourth to the fifth.

The variations in brilliancy that take place in some of the suns that light up distant worlds must have strange consequences. At present one of these suns pours on the planets within his dominion floods of light and heat, and the soil of these planets must remain heated by his burning rays. Some months after, when there is not the smallest cloud in the heavens, the light of the sun begins to wane; it grows more and more ternal; it emits only a pale and wan radiance. The obscuration is gradual, but at last the planet is plunged into the thickest darkness. When the diminution of the sun's light is periodical, the duration of this universal night is for a fixed time, at the end of which the light reappears, unless it is at the end of a term of variation that the darkness is dissipated. The brilliance increases, little by little; and at last we can see the radiant star in all its pristine splendor. Fine weather and brilliant illumination are renewed, until the same waning shall be repeated, and recall the darkness. Can you imagine to what strange vicissitudes Nature must be subjected in regions doomed to alternate torrid heat and icy cold? I am convinced that this glacial period that geologists have established in the history of our globe, — and during which a marked and sudden depression of temperature destroyed a great number of living creatures, and covered Europe with glaciers which descended from the mountains to the plains, — is referable to a momentary diminution in intensity of the Sun's light. Resuming his wonted brilliance, he dissipated the ice that had covered Earth with a mantle of death.*

I have said that there are double stars; that is, worlds lighted by two suns, and sometimes three or four. What is curious in this arrangement is the fact that

---

* See the author's work, "The Earth before the Deluge" (Glacial Period, p. 402, 420).

almost always one of these suns is white, like ours, while the second is colored: it is blue, red, brown, or green. In the constellation Perseus, for instance, it is easy to assure yourself, with a good telescope, that there is a double star. The star η has, indeed, a companion, which belongs to the same solar system: now this second star is blue. This stellar system is at so great a distance that it is impossible for us to measure it with the means used in astronomy for measuring the distances of stars. All that we know, employing the method of optical comparison devised by Sir William Herschel, is that it must take a century for its light to reach us. In the constellation of Ophiuchus there is a similar system of double stars: one is red and the other blue. The same peculiarity is seen in a star of the Dragon. In a double star of Taurus there are visible a red sun and a blue one; and the same collocation occurs in the star η in Argo. There are double solar systems in red and green: such are those of the constellation Hercules, Berenice's Hair, and Cassiopeia. Others are brown and green, and sometimes brown and blue. To this category belong the double stars of the Whale, Eridanus, the Giraffe, Orion, the Unicorn, Gemini, Boötes, and the Swan.

What strange effects must these polychromatic suns produce on the planets that they illuminate! As we know only our Sun, whose light is white, it is difficult for us to imagine the odd consequences that must result from the illumination of a planetary globe and its atmosphere by the rays of blue, red, brown, or green suns.

How queer the soil of these planets must look, the objects that stand on its surface, such as mountains and hills, and the rivers and seas, clouds and vegetation, when all are illuminated by a blue or red light, by floods of scarlet or indigo! We who know Nature in no other guise but that which she wears on the globe in which we are confined can hardly conceive of such effects. What, then, if we could imagine planets lighted during the same day by two successive suns of different colors! It is noon, and a blue sun inundates the globe with floods of its indigo light. The parts strongly illuminated are bright blue, a resplendent azure; those feebly illuminated are dark blue; the half tints are pale blue. Clouds, waters, and vegetation share the common hue. The stars are visible in the day-time, on account of the faint illumination of the heavens. But as the blue sun sinks, see its successor rise on the opposite horizon. It is red, and purple flashes announce its coming. One would think that a mighty conflagration lighted up the east. While on this side of the horizon the purple spreads wider and wider over the heavens, the blue rays gather about the setting sun, and color the curves of the horizon with azure reflections. What a contrast between these two illuminations on the two sides of the heavens! and, in the interval, what strange combinations must result from the fusion of these two lights so diverse in tone! We cannot hope to describe pictures of which nothing around us can suggest even an approximate idea. The poet's imagination and the painter's art would be powerless to conjecture the marvellous effects that the palette of Nature realizes in these enchanted regions.

When two suns, the one red and the other green, or even one brown and the other blue, successively illuminate the same lands, what charming contrasts, what brilliant alternations, must be created by the fusion, which takes place at certain moments, of the red light and the green, or the brown light and the blue! O Nature! what wonderful aspects, what sublime perspectives, thou must put on in those mysterious worlds, to charm the eyes of their fortunate inhabitants!

And the satellites, the moons that light up the nights of their planets, what a strange spectacle must they present, in those strange realms where the feasting of the eyes is eternal! The moon takes on in turn the hues of the two suns, which are reflected, one after the other, on its glowing disk. The phases of the moon seen by the dwellers in these worlds are now red, now blue: hence there is a red quarter of the moon, and a blue quarter. Such a moon has a brown crescent, which succeeds a green one. When it is at the full, the moon of these parts resembles an enormous green fruit wandering in the heavens. There are moons in shades of ruby, detached on the dark ground of the firmament. Others have opaline or azure reflections. Some glitter like diamonds in their circle around the planets which are plunged in shade. O modest moon of ours! no doubt thy peaceful light speaks to our softened and thoughtful souls; but how much deeper must be the impressions, how far more potent the charm, earnest the admiration, and intoxicating the reveries inspired in the dwellers in those far worlds, by the moons of ruby, sapphire, and emerald that illuminate the stillness and serenity of their nights!

THEOPHILUS.

These pictures of the stellar worlds are exceedingly curious, and I thank you for showing them to me. But I do not clearly see your drift, and I fear this long digression, this journey among the stars, has led us away from our subject.

THE AUTHOR.

Not so. After making you understand that the solar system of which we are a part is not unique, that it is but a simple member of a great family of solar worlds, but a little fraction of the Universe, I desired to show, by the diversity of these worlds, the facility with which Nature varies the forces and physical conditions peculiar to the stellar worlds, and consequently the types, animate and inanimate, which compose these worlds. Now, having well grounded you as to the prodigious diversity of the solar worlds that make up the Universe, I come to my main object. I do not lose sight of the fact that my design is here to prove to you that the Universe is boundless; that, by reason of its extent, it is indeed infinity. I approach, then, this great question.

By the mere consideration of the stars, about which you are now so well informed, I am going to bring into relief the immeasurable extent of the Universe. I will speak first of the prodigious distances that separate the stars from the Earth; and the figures will show you that on this side we truly fall into infinity. Next, I will speak of the number of the stars that people space; and on that side, too, the abyss of infinity will yawn before us.

(*Here friend Theophilus grows slightly pale. It is evident that the consideration of infinity disturbs him, as it does every one who for the first time pauses in thought over this unfathomable depth. Nevertheless, he puts on a good face, and listens.*)

THE AUTHOR (*resuming*).

I have, then, at the start, to speak of the distances that divide the stars from the Earth, whence can be logically inferred the distances that separate the same stars from each other. And, first, it should be understood that if, in expressing the distances of the stars, we employed the ordinary unit, — that is, the league of four kilometres, — the figures that it would be necessary to state or write would transcend all limits; and, by the very fact of their inordinate length, would possess no useful significance. For the expression of distances astronomers employ a unit of immense length, and which can consequently serve for the measurement of vast distances. The unit chosen to represent the spaces between the stars is the distance from the Earth to the Sun; in other words, the length of the diameter of the orbit, nearly circular, that the Earth describes in moving around the Sun. The distance from the

Earth to the Sun is thirty-eight millions of leagues, as has already been stated. This distance will be our unit, our standard of measure, in ascertaining the distance of the stars.

I am not sure, by the way, my dear Theophilus, if you have an exact idea of this extent of thirty-eight million leagues that separates us from the Sun. In general, we cannot conceive of distances like those with which astronomy deals, except by representing them by the time occupied by certain moving bodies in the traversing these distances. Resort, then, to comparisons of this kind. A cannon-ball weighing twelve kilogrammes, propelled by six kilogrammes of powder, and moved by a uniform speed of 500 metres per second, would occupy ten years in reaching the Sun from the Earth.* Sound, suppose it to have the same speed in the air as on the surface of the earth, and this speed to be uniform (340 metres per second), would require fifteen years for the same journey. If a railroad through space united the Earth and the Sun, a train of cars travelling at express-train speed, 50 kilometres or twelve and a half leagues per hour, would not reach the terminus till the end of three hundred and thirty-eight years, — leaving the Earth in January, 1872, this supposititious train would not reach the Sun till the year 2210. Let us add that the light of the Sun, which moves at the rate of seventy-seven thousand

* A kilogramme is 2.2055 lbs. avoirdupois; a metre is 3.2808992 feet; a kilometre is 1093.633 yards. — Tr.

leagues per second, must occupy seven minutes thirteen seconds in reaching the Earth.

THEOPHILUS.

So the distance from the Earth to the Sun — that is, thirty-eight millions of leagues — will be our unit of measure for the distances of the stars. Come now to these.

THE AUTHOR.

In order to proceed by very easy stages, I will first take the stars that are nearest to us. Such is one of the stars in the constellation of the Swan. This star is distant from the Earth 551,000 times our unit of measure; that is, we must add to itself 551,000 times the distance from the Earth to the Sun, to represent the distance of the star under consideration, and which, I repeat, is one of the nearest to the Earth. If any one desires to calculate the same distance by the time occupied by its light in its passage, supposing it to travel, like that of our Sun, at the rate of seventy-seven leagues per second, this light would spend nine and a half years in its transit from the star to us.

If you desire now to know the distance of the other stars, — and I do not in all cases, understand, speak of those that are nearest to us, — cast your eyes over this table that I find in a work on Astronomy.

(*Here the Author passes to Theophilus a work on Astronomy, which contains the following table:* —)

DISTANCES OF CERTAIN STARS FROM THE EARTH.

NAMES.	Distances from Earth in terrestrial orbits.	Time of transit of light in traversing this distance.
α of Cygnus	551,000	9½ years.
Vega, α of Lyra	1,330,700	21 „
Sirius, α of Canis Major	1,375,000	22 „
α of Ursa Major	1,550,800	25 „
Arcturus, α of Charles' Wain	1,622,800	26 „
Polar Star	3,078,600	50 „
The Capella, α of Auriga	4,484,000	72 „

THE AUTHOR.

Thus the star Vega (α of the Lyre) is distant from us more than 1,330,000 times the distance from the Earth to the Sun, and its light requires twenty-one years to reach us. The light transmitted to us by the star α of Auriga occupies seventy-two years in its journey. If, by some celestial catastrophe, the star Vega should disappear, be annihilated, we should still see it for twenty-one years, since its light takes that length of time to reach us. As for the star α of the Auriga, it would be visible to us for seventy-two years after its annihilation.

THEOPHILUS.

It may be, then, that our astronomers to-day observe stars that no longer exist, and that are not visible to us, only because the light that they emitted is, at this very moment, travelling toward the Earth.

THE AUTHOR.

Certainly: but to proceed. In order to advance gradually, I have chosen to begin with the star nearest the

Earth. These are stars of the first and second magnitudes. You know, I presume, what the first, second, and third magnitude means, in astronomy?

THEOPHILUS.

Oh, yes! The term magnitude is applied only to the luminous appearance of the star, not to its real volume. It merely indicates its luminous brilliance. A star of the first magnitude is one that belongs to a group of the most luminous stars: one of the second magnitude is that which ranks next in point of brilliance.

THE AUTHOR.

Only note that — a thing sufficiently rare in the exact sciences, and especially in astronomy, the exact science *par excellence* — that here the word "magnitude" signifies just the opposite of what it says. In fact, the more luminous a star seems to us, the nearer it is: on the other hand, the paler and dimmer it is, the farther from us is it. The augmentative terms of first, second, and third magnitudes, &c., then, represent the lustres in a decreasing ratio. The lustre diminishes as the figures grow. Here is an inversion of terms so singular as to be worth noting by the way, and worth remembering for fear of errors.

We have thus far considered only stars of the first and second magnitude. If we come to those of the third, fourth, fifth and sixth magnitude, we shall reach distances quite other than those we have just been treating. These distances are so enormous, reckoning from stars of the third magnitude, that the unit of measure we adopted, immense as it is, — that is, the distance from the Earth to the Sun, — is no longer admissible. The instruments for celestial observation, which can be used for examining and measuring stars of the first and second magnitude, will not serve in dealing with those of the third and larger. By reason of the too small apparent diameter of these stars, which look like mere luminous points, measuring instruments cannot be used. To calculate the distances of stars from the third magnitude, resort is had to a method of comparison based on the amplifying power of telescopes, and which has been successfully employed. I cannot here go into the details of this method, which originated with the astronomer Sir William Herschel, but will content myself with stating its results.

Here, then, are the conclusions reached by this method, as to stars of the sixth magnitude. The light of certain stars of this class would take 1,042 years to reach the Earth : that of others would take 2,700 years. Beyond the sixth magnitude stars are visible only by the aid of the telescope. Of these telescopic stars the distances grow truly stupefying. Some of them are so far from the Earth that their light would be five thousand or even ten thousand years in reaching us, from the moment it left the luminous centre. As to the stars of the last category (the fourteenth magnitude), their light must take a hundred thousand years for its journey to the Earth, supposing that starlight has the same speed as the light of our Sun ; that is, seventy-seven leagues per second.

THEOPHILUS.

But man has existed on the Earth a hundred thousand years, if we may trust the recent works of naturalists; and some of these stars may have been extinguished a hundred thousand years. Mankind, then, must have been able in a hundred thousand years to contemplate those stars which no longer existed. To what strange consequences does a somewhat profound investigation lead us!

THE AUTHOR.

Yes: the luminous rays transmitted to us by these stars lost in the deepest space — that is, telescopic stars — bring us the emanations of solar systems that perhaps no longer exist. The present tells us only the history of the past. I should now have no difficulty in making you understand that there may be stars so profoundly plunged in space, that their light has not yet had time to reach us. They exist, but we cannot see them; not because the telescope is unable to make us discern them, but because the journey of their luminous rays to our globe demands thousands of centuries for its accomplishment, and these thousands of centuries have not yet flowed by. So it is only for our descendants, thousands of centuries hence, that this sight is in store.

I suppose now, friend Theophilus, that you agree with me that the Universe viewed only as to the distances that divide us from the stars, and which must also divide them from each other, is indeed infinity.

THEOPHILUS.

Yes: it is infinity that unrolls before my eyes. Let me breathe a moment.

THE AUTHOR.

You have been contemplating infinity in the distances that separate the stars from the Earth. You will also have a perspective of it, if we consider the number of the stars. When one tries to count these hosts of suns that spangle the heavenly vault as drops of dew spangle the grass of the prairies, one sees that the farther we push into space the more difficult it becomes to enumerate them. If we press farther into these dark gulfs, the suns become so crowded that we must abandon the task, palpably impracticable, of calculating their number. We must leave this sun-dust, this world-seed, to lose itself in the vague darkness of infinite space. Here is what I am going to try to make you conceive, friend Theophilus, so that you may finally pronounce in favor of my reasoning.

I said, just now, that astronomers divided the stars according to their brilliance, in different categories or classes called magnitudes, and which are merely *appearances* dependent on the distance of these stars. I said, also, that our eye ceases to see stars beyond the sixth magnitude, and that the categories of stars above that figure are visible only through the telescope. It is easy to count the stars of the first magnitude; that is, those nearest to us. They are twenty in number.* Those of the second magnitude are sixty-five in number; those of the third, one hundred and seventy. The number of stars increases as their visibility diminishes, and in very rapid proportion.

It has been ascertained that the number of stars in each class of visibility, in apparent magnitude, is three times greater than that of the stars in the preceding class. According to this, we reckon five hundred stars of the fourth magnitude, fifteen hundred of the fifth, and forty-five hundred of the sixth. Summing up the stars in these six categories, which represent the stars

---

* Here is the catalogue of stars of the first magnitude, in the order of their diminishing brilliance: —

1. Sirius, or α in Canis Major.
2. η in Argo.
3. Canopus.
4. α in Centaurus.
5. Arcturus, or α of Charles' Wain.
6. Regel, or β of Orion.
7. Capella.
8. Vega, or α of Lyra.
9. Procyon, or α of Canis Minor.
10. Beteigenze, or α in Orion.
11. Achernar, or α in Eridanus.
12. Aldebaran, or α in Taurus.
13. β in Centaurus.
14. α in the Cross.
15. Antares, or α in Scorpio.
16. Ataïr, or α in Aquila.
17. α in Virgo.
18. α in Pisces Australis.
19. β in the Cross.
20. Pollux.

visible to the naked eye, we reach a total of six thousand. A practised eye, therefore, could succeed in enumerating six thousand stars, in turn, in the two hemispheres and from all parts of the earth.

But the telescope enables us to push very much further the enumeration of the suns: it opens to us the whole depth of the heavens. Instead of the few stars that our eyes can see, it unveils to us a myriad of others like them, so crowded that they seem to cover the sky with a fine silver sand. See, for example (Fig. 17), how a corner of the constellation Gemini looks to the naked eye.

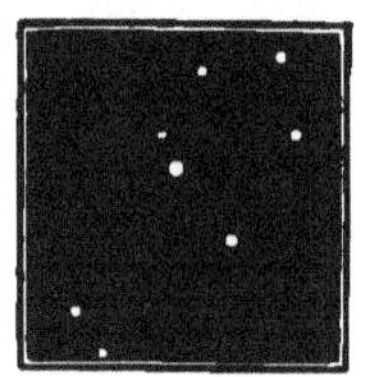

Fig. 17.—A corner of the constellation Gemini.

And see (Fig. 18) how the same part of the heavens appears in the field of the telescope.

It is known that astronomers have succeeded, by the aid of the telescope, in distinguishing, after the sixth magnitude, other successive decreasing magnitudes, and in counting the number of stars belonging to each of these classes. They have distinguished stars of the thirteenth and even of the fourteenth magnitude; that is, stars thirteen or fourteen times more brilliant than Sirius or Vega. The number of stars of the twelfth magnitude is 9,556,000, which, joined to the number of stars in the preceding categories, makes a total, up to the twelfth magnitude, of more than fourteen millions. In the thirteenth magnitude, there are admitted to be, according to the rule which triples the number of stars of one class in order to represent the number in the succeeding class, a total of forty-two millions of stars.

Uniting all the stars visible to the naked eye and those seen by the aid of the telescope, we have then fifty-six millions of suns. And remember, if we pause at this limit, it is because the telescopes now constructed do not enable us to see stars smaller than the thirteenth or fourteenth magnitude. Let these instruments be

Fig. 18.—A corner of the constellation Gemini, seen through the telescope.

still more improved, and we shall see all the regions of heaven covered with this silver sand, this diamond-dust, of which each atom is a sun. And such will be the accumulation of these suns in the depths of space, that we shall see in the field of the telescope nothing but a luminous net-work, formed by the union of suns so close together as to be in apparent contact.

THEOPHILUS.

Come now: this is infinity beginning again. Let me close my eyes.

THE AUTHOR.

Wait a moment. I have not said all: I have only made a beginning. I come now to nebulæ. Among these, you will be afraid of growing dizzy. I suppose you know what astronomers call a nebula?

THEOPHILUS.

Hardly. It is, I think, a diffused, cloudy mass, seen in some parts of the heavens, and whose nature I do not understand. The Milky Way, or the Road to Paradise as it is called in the country, — this, I suppose is what astronomers call nebulæ. It is a certain vaporous mass of luminous matter.

THE AUTHOR.

Yes: you think, as the ancients thought, that a nebula is a collection of vapors that float in space, a kind of cosmic and luminous matter, intended to form worlds some day. You think a nebula is a world in train for formation. Unless you ask yourself, — like the English writer, Derham, author of "Astro-theology," — Is it not the light of the Christian Empyrean, seen through a crack in the heavens? In the eighteenth century the astronomer Halley expressed in scientific terms an error of the same kind, when he wrote, speaking of the nebula of Andromeda: "These spots are simply the light proceeding from a vast space in the region of the ether, a space filled with a diffused and luminous medium." The telescope has put an end to all the hypotheses that could be hazarded as to these celestial appearances, and has also opened new horizons to science as well as to philosophy. Directed upon these luminous masses, the telescope reveals to us their true nature: it has shown that nebulæ are only the union of a very large number of stars. These stars are so numerous that they appear so near to each other as to form a single whole, a single light vague and continuous. But, when their dimensions and distances come to be amplified by the telescope, this diffused light is transformed into a dotted glitter, like that which the heavens present through the same telescope, carpeted with a ground of little stars. A nebula, then, is only the grouping of an enormous number of stars in very close neighborhood. This neighborhood is, however, only apparent. The stars are, in fact, separated from each other by enormous distances. We must not suppose that they are all on the same plane: they belong, on the contrary, to very unequal heights in space; and it is only an optical effect that marshals them on the same apparent plane in the field of the telescope.

One of the nebulosities that give the clearest idea of

these agglomerations of stars is that of Centaurus. This nebula, seen with the naked eye, is but a point in the sky vaguely lighted; but, looked at through a good glass, it takes on the appearance which is shown in Figure 19.

Fig. 19.—Nebula of Centaurus.

An inspection of this figure shows plainly that a nebula does not result from a union of stars merely displayed on a plane of space, but rather of an assemblage of stars situated at unequal distances, and almost making one sphere. In fact, the stars are crowded toward the centre, and more and more distant from each other toward the verge. If we observe from a distance a spherical assemblage of stars, it would have the same aspect,—that is, when we looked at the verge, the visual ray having to traverse but a slight density would encounter but few stars; but, looking at the centre, the visual ray having to pierce the whole density of the mass would encounter many. This leads us to conclude that the nebula of Centaurus, like most agglomerations of this kind, is spherical; that is, that the stars which by their union constitute it are grouped in the form of a sphere.

Is it possible to count the stars that make up a nebula? We cannot go beyond approximations in this direction. Arago, in estimating the intervals between stars situated near the skirts of a nebula, in a position where they did not jut over each other, and then comparing the number of stars at this point with the whole volume of the collection, ascertained that a nebula no larger than a tenth of the apparent disk of the moon contains at least twenty thousand stars. This result gives us an idea of the swarm of suns that the nebulæ enclose; for there are very many of these stellar masses in the heavens.

In the very bosom of the nebulæ, there are luminous points, whose nature the telescope has not yet revealed, and which it has not been able to resolve into stars; but analogy leads us to think that there are still other nebulæ yet more distant, and which escape, by their apparent insignificance, the reach of our instruments. But a day will come when — thanks to the improvements effected in our telescopes — it will be found that these nebulosities are themselves only agglomerations of suns still more remote than the nebula that enfolds them; that is to say, they are, like nebulæ, stellar masses situated in vast depths of space.

The stars that compose the nebulæ are sometimes grouped so as to form regular shapes, — spheres and

Fig. 20. — Nebula of the Crab.

ellipses, more or less elongated. Sometimes the sphere is hollow in the centre, and formed like a link.

Besides these geometric arrangements, there are discovered in nebulæ arrangements altogether irregular and odd. In the constellation Taurus, there is a nebula which, under the telescope, resolves itself into a mass of stars which exhibits the curious arrangement reproduced in Figure 20.

Lord Ross, who first analyzed this nebula, by the aid of the powerful telescope which he had had constructed, gave it the name of the Crab-nebula, because it really resembles that fish in appearance: the antennæ, the legs, and the tail are marked on the dark ground of the sky by a trail of stars.

Fig. 21.—Sobieski's Shield.

Nothing is more varied, nothing more curious, than the forms of the nebulæ which have been studied up to this time, and which exceed a million in number. No two of them are alike.

Some nebulæ seem to be double, or associated. Others trail themselves like serpents, like that of Sobieski's Shield, shown in Figure 21.

Lord Ross first discovered the singular arrangement of the so-called spiral nebulæ. Such a form cannot be explained; but it is certain that the suns which compose the nebulæ are often grouped not around a cen-

Fig 22.—Nebula of Virgo.

tre, not in a shapeless mass, but in perfectly regular curves, according to a system which seems to betray the existence of some mysterious force acting on the stars. The stars are distributed along the lines that represent spirals of different diameters. This is what appears in the nebula of Virgo, as shown in Figure 22.

I said, speaking of the stars, that there are colored stars or suns. I will add here that colored nebulæ are also seen, — red, green, and brown, — which affords another proof of the hypothesis that nebulæ are only agglomerations of stars.

THEOPHILUS.

I see now that the Milky Way is not, as the common people would have it, the Way to Paradise, but a long train of nebulæ.

THE AUTHOR.

There is no doubt of it. This huge, semi-luminous band that crosses the vault of heaven, embracing it like a silver girdle, is not, as has been so long believed, a diffused mass of luminous matter. The telescope, by analyzing the Milky Way, has shown that it consists of a long series of nebulæ. It is an enormous gathering of vastly remote suns, and which we see assembled in an almost regular form, like an oval very much elongated and flattened. The length of the Milky Way is 1,373,000 times the distance from the Earth to the Sun.

THEOPHILUS.

Can we know — can we calculate — the number of stars in the Milky Way?

THE AUTHOR.

There is a paper, by Sir William Herschel, on this question. That observer, who, in the eighteenth century,

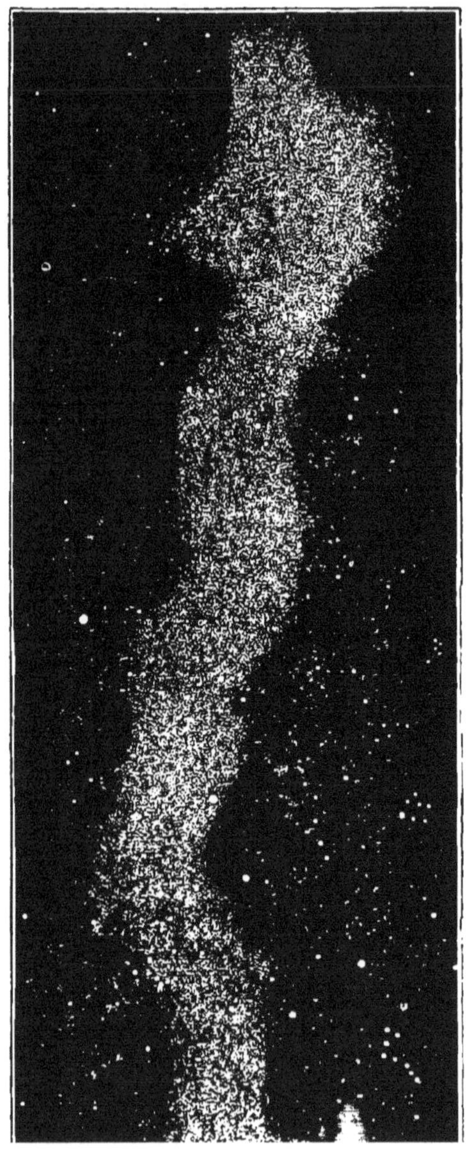

Fig. 23.—The Milky Way.

had constructed the most powerful telescope that had yet been seen, and transported this colossal instrument to the Cape of Good Hope, addressed himself, at that point of observation of the sky of the southern hemisphere, to the task of counting the stars in the Milky Way. Examining only the equatorial part of that huge nebula, he saw pass, in a quarter of an hour, and in a field fifteen minutes in diameter, as many as 116,000 stars: applying this result to the totality of the Milky Way, he found that this nebula must contain more than eighteen million suns. I just said that the length of the Milky Way is 1,373,000 times the Earth's orbit. In order to represent this extent by the time that light takes in passing it, I would say that a ray of light, starting from one of its two extremities, and flying to the other, would occupy fifteen thousand years in its journey. As the Earth and our solar system are in the midst of the Milky Way, it follows that, when we observe with the telescope one of the suns of this nebula, we receive the impression of a luminous ray that left that star seven or eight thousand years ago; that is, before the dawn of historic time.

THEOPHILUS.

I see that we are sailing in absolute infinity.

THE AUTHOR.

If we are not yet, at least we touch it. A last stroke of the oar, and we are in the abyss. Listen further. We have measured the length of the Milky Way, by asserting that light would require 15,000 years to traverse it. This result enables us to measure the extent of other nebulæ still further from us, and in this way. There are sometimes, I have said, in the midst of nebulæ that the telescope resolves into stars, masses of diffused light, which are probably only other nebulæ far more distant. We can even determine the distance of these luminous masses. If it were asked, indeed, to what distance our Milky Way would have to be transported, in order to look like an ordinary nebula (which subtends an angle of about 10′), Arago would answer, according to his investigation on this point, that it would have to be moved three hundred and thirty-four times its length. Now this length is so immense, I said, that it takes light fifteen thousand years to pass from one of its extremities to the other. At 334 times this distance, the nebula of the Milky Way would be seen from the Earth under an angle of 10′; and its light would require for traversing that distance 334 times 15,000, — that is, 5,010,000 years.

Thus the period occupied by light leaving one of these telescopic nebulæ which we mentioned, in reaching us, would be more than *five million years.* Such the distances that may separate from each other the agglomerations of suns suspended in space; such the gaps that exist in the Universe, and that our instruments cannot estimate. It seems to me we are now on the verge of infinity.

THEOPHILUS.

That is evident enough.

THE AUTHOR.

Now, friend Theophilus, when it is understood that these fearful distances that terrify our imagination are merely the results of observations made by our instruments, but that by thought we can increase and extend them, and add them incessantly one to the other; that, to these *five millions of years* of the journey of light, we may in imagination add, if we please, thousands of millions to thousands of millions of years; when we reflect that these innumerable worlds that the telescope reveals to us must be continued still beyond, and even farther and farther; that new agglomerations of suns must succeed these that we can see and measure; that also the suns, the planetary globes and their satellites, add themselves to each other without respite or end, — for the limits assigned to the imagination once reached, a new effort of the mind can press these bounds still farther back, and push even to the deepest depths of unfathomable space this progress towards the dizzy abysses, — then, my dear Theophilus, we understand, as I said in the beginning of this conversation, that the Universe is indeed infinite. And, if you consider that the innumerable battalions of solar systems have each their enforced retinue of planets and satellites, all filled with living beings, — plants, animals, men, and superhumans; if you remember that the flaming comets cross at intervals the orbits of each world, and sink in the fiery furnace of the Sun; that all these thousand millions of suns are infinitely various, that some are double and triple, and some are colored, and pour upon their planets torrents of light, red, blue, green, or brown; and that all these complex movements of these different systems are effected in perfect order, without confusion or collision, — you will see that there is in the Universe not only infinity of extent, but also infinity of order, of harmony, of equilibrium and of law!

THEOPHILUS.

The mind is lost in such thoughts; for the idea of infinity is not adapted to our feeble intelligence. It is a conception forbidden to it. Push no farther, friend, our excursion into this domain in which reason trembles.

THE AUTHOR.

I must, nevertheless, come to the end of my long argument. I must tell you that, in the midst of this boundless space, beyond this enormous company of stars, the homes of living creatures and of feeling souls, there is the Supreme Author, the Sovereign Director, from whom, their sacred spring, proceeds all that our eyes behold, all that our souls feel, all that our intelligence admires, all that my grateful heart blesses; — there is God.

THEOPHILUS.

There you are at the true end of our conversation. And the way has been so long that I think it is time to reach it. The object of this journey through space was to prove that God, being infinite in moral perfections, we must place Him in that infinity of extent that we call the Universe. You have now only to say in what specific spot you fix the home of Divinity, for I do not see what the midst of infinity can be. Having neither beginning nor end, can infinity have a midst?

THE AUTHOR.

I will explain myself on this point. The absolute fixity of the Sun and the stars was a principle in astronomy which, in Newton's time, seemed to be beyond doubt. But Science never pauses. Observations made in this century have shown that the fixity, the immobility of the Sun is only relative. The truth is, the Sun, and with him the whole system of planets, asteroids, satellites, and comets that he leads in his train, change their places. These changes are very slight, no doubt, but are appreciable, and can be measured. Our Sun, with his whole planetary family, seems to tend gently toward that point in the heavens where is the constellation Hercules, and this at the rate of sixty-two millions of leagues per year, or about two leagues per second, which represents one and a half times the radius of the terrestrial orbit. The Sun must describe an orbit that embraces millions of years. But what is true of the Sun must be true of the other suns; that is, of the stars. To the general movement of translation that has been ascertained to exist in our solar system the stellar systems must also be subject; and it is known, beyond doubt, that these thousands of millions of solar systems suspended in boundless space are animated by a motion which bears them, more or less swiftly, toward an unknown point in the heavens. Now nothing tells us that all these circles or ellipses traced by the myriads of solar systems have not a common centre, and that the centre of attraction which our whole solar system obeys in its change of place does not make all the other stars and their systems gravitate to the same point. So all the celestial bodies without exception, the whole swarm of worlds that we have counted, may revolve around the same point, the same centre of attraction. Who now declares that God does not reside in this general focus, this universal centre of attraction of the worlds that fill all space?

THEOPHILUS.

That is the point you desired to reach; and now I comprehend your thought. It impresses me by its grandeur. This God placed in the mathematical centre of the worlds that compose the Universe, this Infinite Intelligence sitting at the centre of the infinite Universe directing the movements of the innumerable hosts of celestial bodies that our imagination can conceive and gather, answers well to the idea that we must form of God, if we dared to face the formidable personality of His Omnipotence. I cannot blame you for recording this theory in your book. It will be in harmony with the quality of religious spirit that animates it, and which, moreover, expresses the desires and aspirations of the men of our time.

There is to-day a strong and deep need of believing

in Providence, of giving homage and faith to God. It is felt that there lies truth, there are peace and safety, now and for ever. But the established religions leave many minds in cruel uncertainties. In the "To-morrow of Death," you have undertaken to lay the foundations of a *religion of Science and Nature.* These principles, I believe, meet the longings of the age. They content the heart and the spirit; they satisfy feeling and reason; they console, they strengthen. In fine, they consecrate the idea of God, without neglecting the Universe or Nature.

THE AUTHOR.

So be it!

www.ingramcontent.com/pod-product-compliance
Lightning Source LLC
Chambersburg PA
CBHW021939220326
41599CB00011BA/827

* 9 7 8 3 3 3 7 4 0 4 5 2 9 *